Smart Devices and Systems for Vibration Sensing and Energy Harvesting

Smart Devices and Systems for Vibration Sensing and Energy Harvesting

Editors

Kai Tao
Yunjia Li

MDPI • Basel • Beijing • Wuhan • Barcelona • Belgrade • Manchester • Tokyo • Cluj • Tianjin

MDPI

Editors

Kai Tao
Department of Microsystem Engineering
Northwestern Polytechnical University
Xi'an
China

Yunjia Li
Electrical Engineering
Xi'an Jiaotong University
Xi'an
China

Editorial Office
MDPI
St. Alban-Anlage 66
4052 Basel, Switzerland

This is a reprint of articles from the Special Issue published online in the open access journal *Micromachines* (ISSN 2072-666X) (available at: www.mdpi.com/journal/micromachines/special_issues/smart_devices_systems).

For citation purposes, cite each article independently as indicated on the article page online and as indicated below:

LastName, A.A.; LastName, B.B.; LastName, C.C. Article Title. *Journal Name* **Year**, *Volume Number*, Page Range.

ISBN 978-3-0365-6835-5 (Hbk)
ISBN 978-3-0365-6834-8 (PDF)

Contents

micromachines

Editorial

Editorial for the Special Issue on Smart Devices and Systems for Vibration Sensing and Energy Harvesting

Kai Tao [1,2,*] and Yunjia Li [3,*]

1 Ministry of Education Key Laboratory of Micro and Nano Systems for Aerospace, Northwestern Polytechnical University, Xi'an 710072, China
2 Research and Development Institute in Shenzhen, Northwestern Polytechnical University, Shenzhen 518057, China
3 School of Electrical Engineering, Xi'an Jiaotong University, Xi'an 710049, China
* Correspondence: taokai@nwpu.edu.cn (K.T.); liyunjia@xjtu.edu.cn (Y.L.)

The Internet of things (IoT) poses new challenges for sensors and their power systems [1,2]. The deployment of large numbers of sensor nodes requires the sensor to work for a sufficient period without battery replacement. As self-sustained power sources, micro/nano energy harvesting systems can capture and transform unused ambient energy into electrical energy [3–5]. They have been regarded as an alternative to conventional electrochemical batteries, which will pave the way for actualizing energy-autonomous devices and intelligent monitoring activities. Integrating micro/nano power sources with IoT will be a revolutionary technology in the following decades. In addition, as many of the vibration energy harvester devices are sensitive to vibration, they are inherently considered as excellent candidates for vibration sensing.

This Special Issue focuses on state-of-the-art vibration sensing and energy harvesting technologies. After being carefully reviewed, 11 articles have been accepted for publication in this Special Issue [6–16] that can be divided into three aspects: (i) new materials for vibration energy harvesting applications, (ii) design and fabrication of vibration energy harvesters (VEHs), and (iii) system-level integration and testing of VEHs.

New materials are crucial for breaking the bottleneck in the performance of the VEHs. Zhang et al. investigated the influence of ferromagnetic material on the output characteristics of the Halbach array energy-harvesting structure. After adding the iron sheet, the energy-harvesting efficiency of the Halbach array can be significantly improved [6]. Chen et al. proposed a novel double-network ionic hydrogel and fluorinated ethylene propylene (FEP) electret-based tactile sensor. Combining with a corona-charged FEP film, the output performance of the sensor is significantly boosted by 156.3% through the hybrid of triboelectric and electrostatic effects [7].

Structural design is another essential aspect for boosting the output performance of the VEHs. A torsional oscillating magnet-based vibration energy harvester was reported by Wang et al. Microfabricated silicon torsional springs contributed to the torsional movement of the magnet, which effectively reduced the footprint of the device and achieved an output power of 6.9 μW with a size of 1 cm × 1 cm × 1.08 cm [8]. Li et al. proposed an electromagnetic vibration energy harvester with tunable resonance frequency based on the stress modulation of flexible springs. By physically pulling and pushing the springs via a pair of metallic fixtures, the total tuning frequency range is 56 Hz (74–130 Hz) [9]. To improve the energy harvesting performance of an energy harvester, Chen et al. proposed a novel bistable piezoelectric energy harvester by introducing a linear-arch beam, the dynamics model of the system was established, and the simulation was conducted to verify the feasibility [10]. Furthermore, Zhang et al. reported a tri-stable piezoelectric energy harvester based on a linear-arch composite beam. The experiment result shows that the system can be mono-stable, bi-stable, and tri-stable by adjusting the horizontal or vertical

Citation: Tao, K.; Li, Y. Editorial for the Special Issue on Smart Devices and Systems for Vibration Sensing and Energy Harvesting. *Micromachines* **2023**, *14*, 173. https://doi.org/10.3390/mi14010173

Received: 5 January 2023
Accepted: 5 January 2023
Published: 10 January 2023

spacing of the magnets [11]. Xu et al. utilized the vibration energy to drive a single-phase ultrasonic motor directly. The motor obtains advantages in structure miniaturization and circuit simplification [12]. Dong et al. reported stretchable strain sensors with a controllable negative resistance sensitivity coefficient by designing a silica gel/CNTs/silica gel sandwich structure [13].

Optimization of the system-level integration methodologies of VEHs is important for their practical applications. Cao et al. proposed an electrostatic-piezoelectric-electromagnetic hybrid vibrational power generator. By designing the energy management circuit at the rear end of the device, the packaged device can directly export a 3.3 VDC voltage to supply power to most of the sensing equipment [14]. A novel hydraulic interconnected regenerative suspension was reported by Guo et al. [15]. The root means square values of the bounce and roll acceleration of the proposed system are 64.62% and 11.21% lower than that of a standard suspension, respectively. The suspension system could output 186.93, 417.40, and 655.90 W at speeds of 36, 72, and 108 km/h for an off-road vehicle on a Class-C road, respectively. Apart from the design and fabrication processes, the reliability study on the VEHs is also necessary for the application of vibration energy harvesters. For this purpose, Li et al. presented a micro-tester for measuring the reliability of the energy harvester and other MEMS devices in controlled environmental conditions [16]. The system could generate a temperature range of 0–120 °C and a humidity range of 20–90% RH (0–55 °C) within a small footprint and weight.

In conclusion, the researchers reported large numbers of innovative research progresses in materials, device configuration, and system-level VEHs. This Special Issue gathers the latest developments in smart devices and systems for vibration sensing and energy harvesting applications, which could serve as a good reference for VEH research.

Conflicts of Interest: The authors declare no conflict of interest.

References

1. Li, Y.J.; Tao, K.; George, B.; Tan, Z.C. Harvesting Vibration Energy: Technologies and Challenges. *IEEE Ind. Electron. Mag.* **2021**, *15*, 30–39. [CrossRef]
2. Tao, K.; Wu, J.; Tang, L.; Xia, X.; Lye, S.W.; Miao, J.; Hu, X. A novel two-degree-of-freedom MEMS electromagnetic vibration energy harvester. *J. Micromech. Microeng.* **2016**, *26*, 035020. [CrossRef]
3. Yang, Z.B.; Zhou, S.X.; Zu, J.; Inman, D. High-Performance Piezoelectric Energy Harvesters and Their Applications. *Joule* **2018**, *2*, 642–697. [CrossRef]
4. Suzuki, Y. Recent Progress in MEMS Electret Generator for Energy Harvesting. *IEEJ Trans. Electr. Electron. Eng.* **2011**, *6*, 101–111. [CrossRef]
5. Wang, Z.L. Triboelectric Nanogenerators as New Energy Technology for Self-Powered Systems and as Active Mechanical and Chemical Sensors. *ACS Nano* **2013**, *7*, 9533–9557. [CrossRef] [PubMed]
6. Zhang, X.; Liu, H.; He, Y.; Peng, T.; Su, B.; Guan, H. Analysis of the Influence of Ferromagnetic Material on the Output Characteristics of Halbach Array Energy-Harvesting Structure. *Micromachines* **2021**, *12*, 1541. [CrossRef] [PubMed]
7. Chen, Z.; Yu, J.; Zeng, H.; Chen, Z.; Tao, K.; Wu, J.; Li, Y. An Electret/Hydrogel-Based Tactile Sensor Boosted by Micro-Patterned and Electrostatic Promoting Methods with Flexibility and Wide-Temperature Tolerance. *Micromachines* **2021**, *12*, 1462. [CrossRef] [PubMed]
8. Wang, X.; Li, J.; Zhou, C.; Tao, K.; Qiao, D.; Li, Y. Vibration Energy Harvester Based on Torsionally Oscillating Magnet. *Micromachines* **2021**, *12*, 1545. [CrossRef] [PubMed]
9. Li, Y.; Zhou, C.; Cao, Q.; Wang, X.; Qiao, D.; Tao, K. Electromagnetic Vibration Energy Harvester with Tunable Resonance Frequency Based on Stress Modulation of Flexible Springs. *Micromachines* **2021**, *12*, 1130. [CrossRef] [PubMed]
10. Chen, X.; Zhang, X.; Chen, L.; Guo, Y.; Zhu, F. A Curve-Shaped Beam Bistable Piezoelectric Energy Harvester with Variable Potential Well: Modeling and Numerical Simulation. *Micromachines* **2021**, *12*, 995. [CrossRef] [PubMed]
11. Zhang, X.; Chen, L.; Chen, X.; Zhu, F.; Guo, Y. Time-Domain Dynamic Characteristics Analysis and Experimental Research of Tri-Stable Piezoelectric Energy Harvester. *Micromachines* **2021**, *12*, 1045. [CrossRef] [PubMed]
12. Xu, D.; Yang, W.; Zhang, X.; Yu, S. Design and Performance Evaluation of a Single-Phase Driven Ultrasonic Motor Using Bending-Bending Vibrations. *Micromachines* **2021**, *12*, 853. [CrossRef] [PubMed]
13. Dong, R.; Xie, J. Stretchable Strain Sensor with Controllable Negative Resistance Sensitivity Coefficient Based on Patterned Carbon Nanotubes/Silicone Rubber Composites. *Micromachines* **2021**, *12*, 716. [CrossRef]
14. Cao, Y.; Shen, W.; Li, F.; Qi, H.; Wang, J.; Mao, J.; Yang, Y.; Tao, K. All-in-One High-Power-Density Vibrational Energy Harvester with Impact-Induced Frequency Broadening Mechanisms. *Micromachines* **2021**, *12*, 1083. [CrossRef] [PubMed]

15. Guo, S.; Chen, L.; Wang, X.; Zou, J.; Hu, S. Hydraulic Integrated Interconnected Regenerative Suspension: Modeling and Characteristics Analysis. *Micromachines* **2021**, *12*, 733. [CrossRef] [PubMed]
16. Li, Y.; Dou, W.; Zhou, C.; Wang, X.; Yang, A.; Zhang, Y.; Qiao, D. A Microtester for Measuring the Reliability of Microdevices in Controlled Environmental Conditions. *Micromachines* **2021**, *12*, 585. [CrossRef] [PubMed]

micromachines

Article

An Electret/Hydrogel-Based Tactile Sensor Boosted by Micro-Patterned and Electrostatic Promoting Methods with Flexibility and Wide-Temperature Tolerance

Zhensheng Chen [1], Jiahao Yu [1], Haozhe Zeng [1], Zhao Chen [1], Kai Tao [1,*], Jin Wu [2,*] and Yunjia Li [3,*]

[1] Ministry of Education Key Laboratory of Micro and Nano Systems for Aerospace, Northwestern Polytechnical University, Xi'an 710072, China; 2019201174@mail.nwpu.edu.cn (Z.C.); 1036597329@mail.nwpu.edu.cn (J.Y.); 2018300323@mail.nwpu.edu.cn (H.Z.); chenzhao1999@mail.nwpu.edu.cn (Z.C.)

[2] State Key Laboratory of Optoelectronic Materials and Technologies and the Guangdong Province Key Laboratory of Display Material and Technology, School of Electronics and Information Technology, Sun Yat-sen University, Guangzhou 510275, China

[3] School of Electrical Engineering, Xi'an Jiaotong University, Xi'an 710049, China

* Correspondence: taokai@nwpu.edu.cn (K.T.); wujin8@mail.sysu.edu.cn (J.W.); liyunjia@xjtu.edu.cn (Y.L.)

Abstract: With the rising demand for wearable, multifunctional, and flexible electronics, plenty of efforts aiming at wearable devices have been devoted to designing sensors with greater efficiency, wide environment tolerance, and good sustainability. Herein, a thin film of double-network ionic hydrogel with a solution replacement treatment method is fabricated, which not only possesses excellent stretchability (>1100%) and good transparency (>80%), but also maintains a wide application temperature range (−10~40 °C). Moreover, the hydrogel membrane further acts as both the flexible electrode and a triboelectric layer, with a larger friction area achieved through a micro-structure pattern method. Combining this with a corona-charged fluorinated ethylene propylene (FEP) film, an electret/hydrogel-based tactile sensor (EHTS) is designed and fabricated. The output performance of the EHTS is effectively boosted by 156.3% through the hybrid of triboelectric and electrostatic effects, which achieves the open-circuit peak voltage of 12.5 V, short-circuit current of 0.5 μA, and considerable power of 4.3 μW respectively, with a mentionable size of 10 mm × 10 mm × 0.9 mm. The EHTS also demonstrates a stable output characteristic within a wide range of temperature tolerance from −10 to approximately 40 °C and can be further integrated into a mask for human breath monitoring, which could provide for a reliable healthcare service during the COVID-19 pandemic. In general, the EHTS shows excellent potential in the fields of healthcare devices and wearable electronics.

Keywords: flexible electronics; electret/hydrogel-based tactile sensors; pyramidal parented hydrogel; anti-freezing and anti-drying

Citation: Chen, Z.; Yu, J.; Zeng, H.; Chen, Z.; Tao, K.; Wu, J.; Li, Y. An Electret/Hydrogel-Based Tactile Sensor Boosted by Micro-Patterned and Electrostatic Promoting Methods with Flexibility and Wide-Temperature Tolerance. *Micromachines* **2021**, *12*, 1462. https://doi.org/10.3390/mi12121462

Academic Editor: Ai Qun Liu

Received: 3 October 2021
Accepted: 25 November 2021
Published: 27 November 2021

1. Introduction

The rapid development of wearable and functional electronics has drawn significant attention towards the field of transparent, flexible, and efficient devices that are urgently demanded by human beings [1–7]. Various devices and applications of mechanical sensors [2,8–10], flexible electronic skins [11–15], and wearable devices [16–18] have been designed and introduced into our lives, which facilitate the evolution of human science and technology. Various flexible devices have been widely employed and investigated for their merits of easy integration, outstanding biocompatibility, and mechanical characteristics [19–26]. The conventional power supply methods for the devices mentioned above, such as batteries, unavoidably lead to environmental problems. Additionally, the poor flexibility and inconvenient maintainability of traditional devices give rise to severe restrictions for their biophysical applications in wearable electronic device industry. Hence, the capability of harvesting energy from external stimulations, namely self-powered, to drive the devices or generate output signals, is urgently demanded [27–34].

The triboelectric effect has received enormous interest for its effective power generation function and was first proposed by Prof. Zhonglin Wang in 2012 as triboelectric nanogenerators (TENGs) [4]. Subsequently, many absorbing devices have been presented and studied, including TENGs with intense output power density, entirely newly designed structures for ocean energy harvesting, ultra-thin devices for wearable biophysiological sensing, and integration in diverse engineering applications [35–43]. Due to the synergistic effect of triboelectrification and electrostatic induction, TENG will generate the output signals continuously according to the external stimulations. When two friction layers with different electron affinities contact and separate with each other, a redistribution of charges happens in the external circuit, which will generate an alternating current responding to the dynamic variations of the stimulations, capable of powering small electronic devices [44].

However, there still remain two significant drawbacks that need to be considered for the self-powered electronics. The first problem is that most flexible TENGs' performances are limited by environmental temperatures. Traditional electrodes utilized in traditional TENGs are generally metals, which lead to a rigid and opaque device, further restricting their applications seriously [45–49]. Some flexible electrodes, such as liquid metals or elastomeric polymers, are currently adopted by TENGs [11,50–53], and such materials improve the flexibility and deformable ability a lot but are also constrained by their poor temperature tolerance, resulting in the invalidation phenomenon in extreme environments [54–59]. To solve this problem, a double-networked ionic hydrogel with a solvent replacement treatment method is proposed and applied in this work. Due to the excellent stretchability, electrical stability, and wide temperature tolerance of the hydrogel [60], the EHTS device in this paper demonstrates great potential in the field of wearable and flexible electronics under some extreme environments.

The second problem here is that the output performance of most self-powered sensors is primarily constrained by the efficiency of triboelectrification and electrostatic induction, which naturally depends on the materials utilized. To increase the output signals, a micro-pyramid patterned spin coating method is introduced in the hydrogel fabrication process to further enlarge the triboelectric(contacting) area, and thus the output voltage increases significantly. Moreover, a fluorinated ethylene propylene (FEP) film with an effective corona discharging processing method is integrated into the EHTS, which further promotes performance.

In this article, we proposed an electret/hydrogel-based tactile sensor (EHTS) with a pyramidal patterned double-networked (DN) ionic hydrogel membrane and charged FEP electret thin film to realize self-sustained physical detection (Figure 1a). As is implemented in our previous work [2], conductive silver nanowires (AgNWs) synthesized with polyethylene terephthalate (PET) film were adopted as the flexible substrate for the AgNWs electrode. The fabricated EHTS has a basic cavity structure which consists of an FEP and AgNWs film opposite the micro-pyramidal structured DN ionic hydrogel film. During the simple operating mode of pressing and releasing, both triboelectric and electrostatic effects occured to generate the external currents between the top AgNWs and the bottom hydrogel electrodes (Figure 1b(i,ii)). Due to the high transparency and deformability of hydrogel and FEP and AgNWs layers, EHTS demonstrates excellent visibility and flexibility (As is demonstrated in Figure 1c, ETHS could be easily adhered to the fingertip. Inset image shows the size of the EHTS on top of a fifty-cent coin). The scanning electron microscope (SEM) images of the nanostructured FEP film (Figure 1d) and AgNWs electrode (Figure 1e) show the roughness surface topography, which boosts the ability to capture more charges to improve the performance of EHTS. Besides, to further increase the contact area to enhance the output performance of EHTS, micro-pyramidal patterned hydrogel (confocal laser scanning microscope image of Figure 1f) was employed to promote the triboelectric effect of the device.

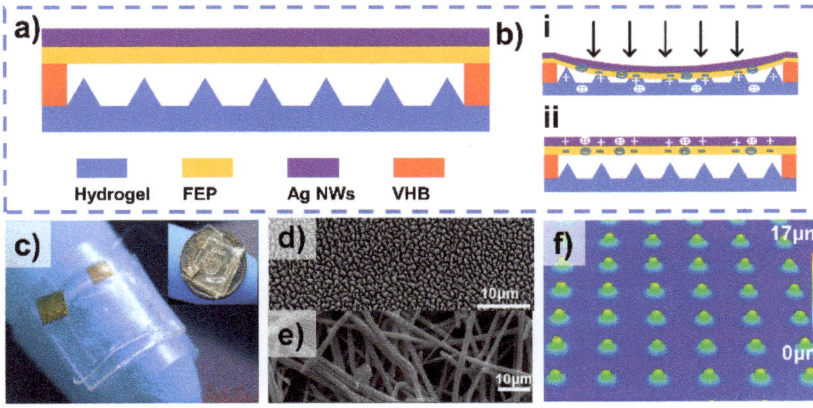

Figure 1. The basic structure of (electret/hydrogel-based tactile sensor EHTS): (**a**) Structure design of the proposed EHTS; (**b**) Schematic diagram of the ETHS at the compressed state (i) and released state (ii); (**c**) Photograph of the fabricated device attached on the fingertip. The inset is the ETHS on a fifty-cent coin; (**d**,**e**) Scanning electron microscopy (SEM) images of the nanostructured fluorinated ethylene propylene (FEP) film surface and the fabricated AgNWs electrode; (**f**) Confocal Laser Scanning Microscope (CLSM) image of the micro-pyramidal structures on the hydrogel.

2. Materials, Characteristics and Fabrication Methods

Traditional DN hydrogel is often supposed to be a promising material for flexible and transparent electronics due to its ultra-high deformability, outstanding transparency, excellent mechanical strength, and ionic conductivity. However, the above properties of DN hydrogel can be seriously affected by the external temperature, which is mainly due to the large water content of the hydrogel (>80% usually). Thus, the performance of hydrogel-based electronics is often constrained by their low-temperature tolerance, which leads to severe limitations of their practical value. Herein, a novel solvent replacement treatment (SRT) based on a saturated solution of LiBr is utilized in the hydrogel fabrication process to improve the temperature tolerance. To investigate the wide-temperature tolerance of the SRT hydrogel, comparison tests between the pristine hydrogel and the hydrogel with SRT are presented. A carrageenan/polyacrylamide (PAM) DN hydrogel is utilized in this work as the pristine hydrogel by a simple thermal polymerization method. The PAM and carrageenan can be naturally cross-linked via covalent bonds and particle bonds (As is schematically demonstrated in Figure 2a). The manufactured DN hydrogel is cut into five samples with the same size of 20 mm × 5 mm × 5 mm to investigate the environmental tolerance properties between the pristine hydrogel and the hydrogel with SRT. Two samples are immersed in the 50 wt% LiBr solution (saturated solution at room temperature) to fabricate the hydrogel samples with SRT while the others stay untreated. The pristine DN hydrogel sample shows excellent flexibility and transparency as shown in Figure 2a. However, free water molecules in the pristine hydrogel remain unbonded, which will freeze or evaporate naturally due to the temperature change in the external environment. After being kept in the oven (set for 40 °C) for 1 h, the pristine hydrogel shrunk severely due to the rapid evaporation of water (shown in Figure 2b), which results in the severe dehydration and significant rigidity of the test sample (inset in Figure 2b). On the contrary, the counterpart with SRT remains unchanged under the same condition due to the formation of ion complexes between the LiBr and water molecules (right inset of Figure 2d), which leads to the strong ionic hydration effect of the hydrogel with SRT (Figure 2d). For the extreme cold condition, Figure 2c schematically illustrates the phase transformation of the free water molecules in the pristine hydrogel. After being stored at −10 °C in a refrigerator for 1 h, a test sample of pristine hydrogel thoroughly froze (inset in Figure 2c). Conversely, owing to the low freezing point of LiBr solution, no icing phenomenon is found in the DN ionic hydrogel sample with SRT at the same experimental conditions.

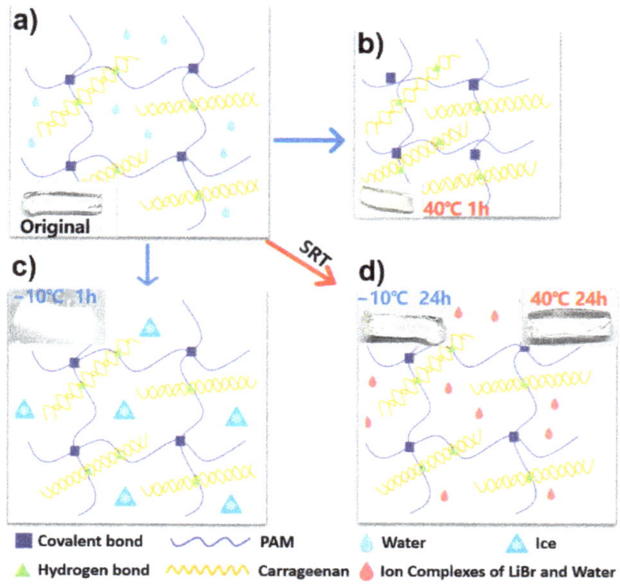

Figure 2. (**a**) Schematic illustration of the double-networked (DN) hydrogel without solvent replacement treatment (SRT) at original state. Inset is the photograph of the original hydrogel; (**b**,**c**) Dehydration and freeze state of the original DN hydrogel when stored at 40 °C and −10 °C for 1 h; (**d**) Anti-freezing and anti-drying properties of DN hydrogel after SRT.

Due to the strongly interconnected networks of carrageenan and PAM chains as well as the covalent bonds and hydrogen bonds in the DN ionic hydrogel (as is demonstrated in Figure 2d), excellent stretchability and flexibility can be achieved. The stretch test of the fabricated DN ionic hydrogel is implemented in Figure 3a. The original hydrogel sample (cut to 7.5 mm × 4 mm × 3 mm) can be easily stretched over 1180% in a mechanical tensile platform (Figure 3a(i–iii)), demonstrating the promising ductility of the DN ionic hydrogel. By a simple spin coating method, DN ionic hydrogel with SRT (hereinafter referred to as SRT hydrogel) achieves ultra-thin thickness within 200 μm. Therefore, the fabricated SRT hydrogel membrane can be easily twisted, rolled, and folded (Figure 3b(i–iii)), which contributes a lot to the excellent flexibility as well as the considerable transparency of the fabricated EHTS. As the transmittance testing results shown in Figure 4a, the SRT hydrogel membrane and EHTS remain around 90% and 80% luminousness in the range of 400 nm~800 nm (visible region), respectively. A logo beneath the SRT hydrogel membrane can be observed clearly (as the inset of Figure 4a), which could bring tremendous application potential for EHTS in wearable electronics.

Additional experiments have been carried out to evaluate the resistance performance while stretching the SRT hydrogel (Figure 4b). The dynamic resistance variation shows the excellent electrical stability of the SRT hydrogel sample in the cyclic test (10 cycles for 50%, 100%, 150%, and 200% strain state, respectively). Additionally, the resistance of SRT hydrogel is positively correlated with the stretching state. As the tensile strain increases from 0% to 335% sequentially, resistance changes accordingly. The same trend can also be found in the releasing process (Figure 4c). Therefore, we chose SRT hydrogel as the flexible electrode as well as the triboelectric layer in EHTS.

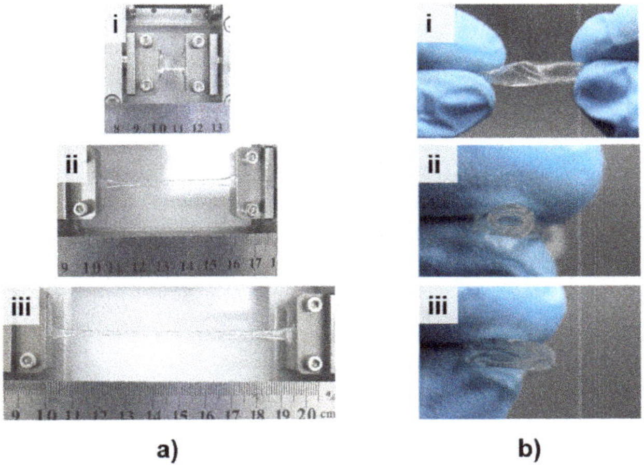

Figure 3. (**a**) Elongation test of the fabricated DN ionic hydrogel at 0% (i), 750% (ii) and 1180% (iii); (**b**) Photographs of the DN ionic hydrogel membrane at twisting (i), rolling (ii) and folding (iii) state.

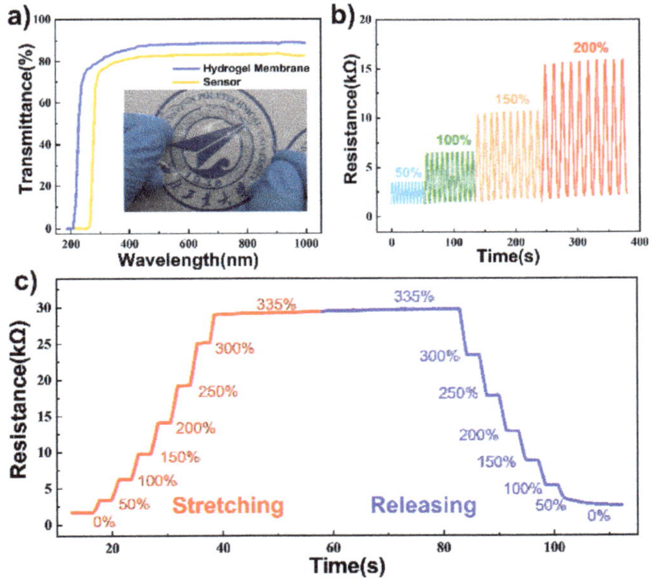

Figure 4. (**a**) Transmittances of the fabricated DN ionic hydrogel membrane and the EHTS; (**b**) Resistance variation of the hydrogel under different cyclic stretching states; (**c**) Real-time response of the hydrogel resistance when stretched to 335% and then released.

To fabricate EHTS, a production process has been put forward, including a series methods of Si wafer etching, spin coating, peeling off, FEP discharging, and final assembly (as shown in Figure 5). The Si wafer (100) was oxidized in a furnace to grow a SiO_2 layer of 300 nm thickness on the substrate surface. The thermally oxidized Si wafer is spin-coated by the Shipley1805 (MicroChem, Round Rock, TX, USA) photoresist (PR) (Figure 5a). As is demonstrated in Figure 5b, the Cr-coated hard photomask was utilized to fabricate the blank square array on PR layer (by SUSS MJB4 UV400, SUSS MicroTec company, Garching, Germany). After the exposure process of PR, the wafer was immersed in the buffered oxide etch (BOE, NH4F/HF = 7:1, *v*/*v*, Transene Company, MA, USA) for 4 min to remove the SiO_2 under the square array (Figure 5c). Then, the PR was removed by sonication in acetone (Figure 5d). The anisotropy of the wet etching by KOH etching solution (30% KOH in H_2O/isopropanol (4:1 v/v)) was carried out for 2 mins, ensuring to format the inverted pyramidal structures (with the) on the Si surface (Figure 5e). To complete the fabrication of the Si mold, the SiO_2 layer was removed through BOE solution treatment (Figure 5f). The micro-pyramid patterned DN hydrogel membrane can be achieved by spin coating the pre-mixed hydrogel solution at the speed of 800 rpm for 15 s (Figure 5g). After the formation, the DN hydrogel membrane is peeled off, followed with the SRT to acquire the SRT hydrogel layer (Figure 5h). As shown in Figure 5i, the AgNWs/FEP composite is prepared as the triboelectric layer and flexible electrode by a typical process method described in our previous research. The fabricated AgNWs/FEP film is charged by the corona discharging system (Figure 5j). Finally, the formed micro-patterned SRT hydrogel is assembled with the charged AgNWs/FEP film with double-sided tape to ensure the charged FEP film is opposite the micro-patterned hydrogel surface (Figure 5k). At last, EHTS is completely fabricated by connecting the hydrogel and AgNWs electrodes together.

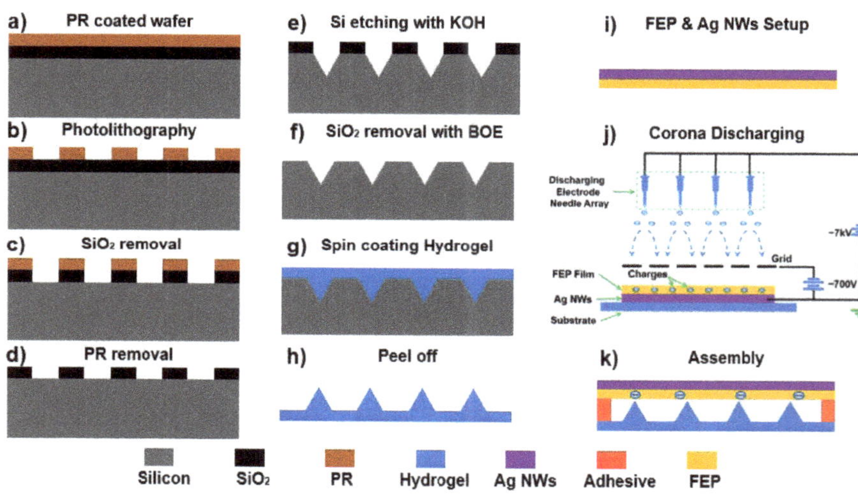

Figure 5. Fabrication of the micro-pyramidal patterned hydrogel membrane and EHTS. (**a**) The thermally oxidized Si wafer was spin-coated with Shipley1805 photoresist (PR); (**b**) PR was partial removed through photolithography; (**c**) The SiO_2 was etched in BOE plasma; (**d**) PR was completely removed; (**e**) Si etching with KOH to form the inverted pyramids array; (**f**) SiO_2 layer was removed with BOE thoroughly; (**g**,**h**) The micro-pyramidal patterned hydrogel film was spin-coated and peeled off; (**i**) FEP electret film and AgNWs electrode are bonded together with ethoxylate resin; (**j**) Corona discharging method is adopted to implant charge into FEP film; (**k**) EHTS was fabricated by assembled micro-pyramidal patterned hydrogel film and charged FEP & AgNWs together.

3. Working Principles

Figure 6 shows the charge circulation and the electric field variations of the fabricated EHTS in the compress–release cycle. Figure 6a demonstrates the initial state without any stimulation, when charged FEP/AgNWs film and hydrogel layer are opposite with, and separate from each other. After the external force is applied on EHTS, the gap distance between two plates decreases and electrostatic induction happens naturally. The implanted electrons in FEP film induce the positive charges transfer through the external circuit to generate the current (Figure 6b). As the deformation intensifies, the triboelectrification effect occurs when the FEP film contacts the hydrogel surface, which will further increase the output current (Figure 6c). After the compressing force is removed, EHTS will recover to its original state and drive the charges flow back to the balanced state again (Figure 6d). During this cycle, both electret-based electrostatic and triboelectric effects work together to magnify the output performance of EHTS. Moreover, due to the deformation of the micro-pyramids on the hydrogel surface, the contact area is increased, which leads to a stronger triboelectrification effect. Numerical simulations of EHTS have also been calculated by Comsol Multiphysics software (Version 5.5, COMSOL, Inc., Stockholm, Sweden) to analyze the potential variations. Figure 6e shows the initial potential distribution of EHTS at the original state. The upper FEP film has already been corona charged, which provided the strong negative potential. As the applied force increases, the distance between FEP film and hydrogel layer becomes closer due to the compressing deformation of the FEP & Ag NWs layers, which causes electrostatic induction and the variation of the potential (Figure 6f). Eventually, with the increase of the applied pressure, FEP film fully contacted the bottom hydrogel. The different electron affinities between the hydrogel electrode and FEP film lead to the triboelectric effect in the contact interface. The potential difference between the two friction layers has been dramatically increased (Figure 6g). Thus, the output performance of EHTS has been boosted significantly by the combination promoting efforts of the triboelectric and electrostatic effects. As a result, potential distribution varies accordingly when two contacted layers separate to their original states (Figure 6h).

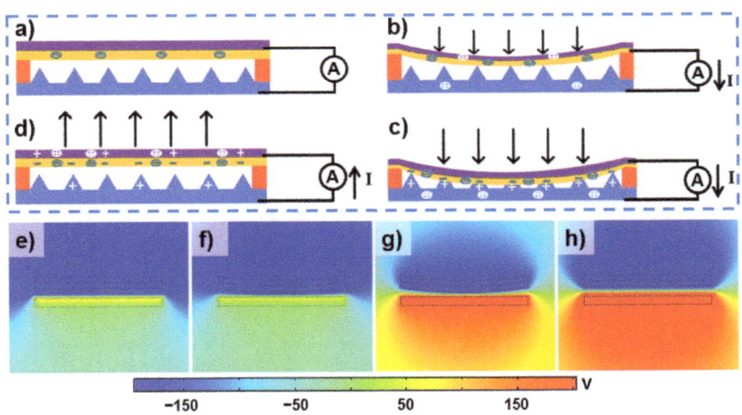

Figure 6. Working principle of the EHTS: Original (**a**), Compressing (**b**), Compressed (**c**) and recovered (**d**) state; Electric field variation of EHTS during the compress-recovery cycle (**e**–**h**).

4. Results

The enhanced effects of the output performance obtained by micro-pyramidal patterned method and charged FEP electret have been further investigated. Four kinds of EHTSs are fabricated by different process methods, such as the EHTS with plain hydrogel and non-charged FEP, the device with micro-pyramidal patterned hydrogel but non-charged FEP, the plain hydrogel with charged FEP, and the EHTS with both microstructures and charged FEP. The output voltages of the three EHTSs are measured at the same stimulation conditions by a vibration platform at 1 Hz. As we can obviously conclude in Figure 7a, the 1st, 2nd, 3rd, and 4th EHTS outputs the signals with the average peak voltage of 4.8 V, 8.6 V, 9.8 V, and 12.3 V, respectively. Therefore, the enhancement rates of the micro-pyramidal patterned method and charged FEP film are about 79% and 104%. With the combined effect of microstructures and FEP electret, EHTS generates a considerable peak open-circuit voltage of 12.5 V and an impressive short-circuit current of 0.54 μA (Figure 7b). The maximum output voltage of the experiments result is approximately only 1/10 of the simulation one. The output voltage is related to the charge density of the electret according to the relative research implemented by X. Ye, et al. [31]. During negative corona charging, a lot of positive charges are injected into the back-side of the hydrogel surface, which indicate the double sides surface potential are pretty close. Therefore, difference between the front and back surface charge densities is much smaller than the charge density of the front side, which is identified as the main reason for the much lower result. As is shown in Figure 7c, the maximum instantaneous output power of EHTS achieves 4.3 μW with an optimum load resistance of 24 MΩ. What needs to be emphasized is that all the above output performances are generated by the EHTS with the small size of 10 mm × 10 mm × 0.9 mm, which is sufficient to be integrated in wearable electronics and sensors with a strict dimension limitation.

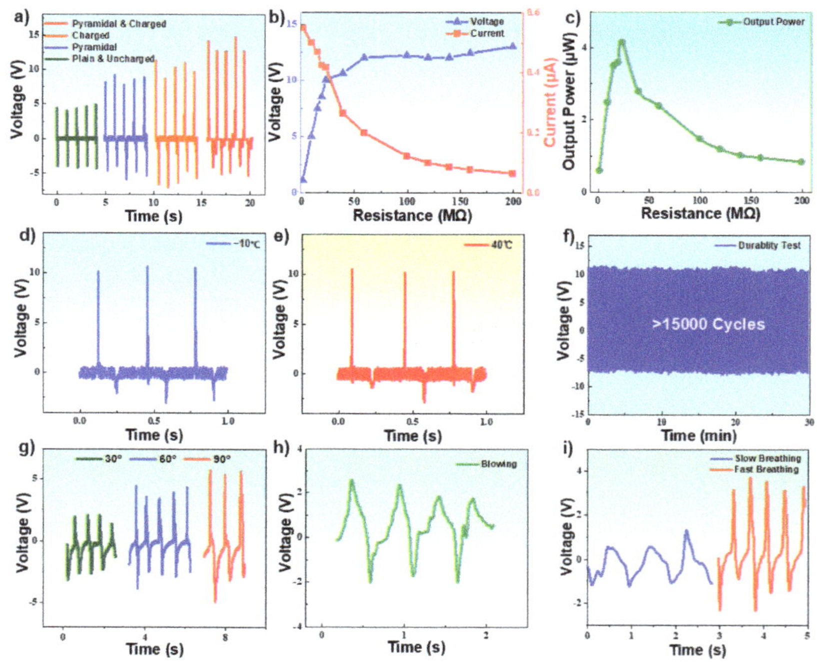

Figure 7. (**a**) Comparison of the output performance between plain & uncharged, pyramidal hydrogel & non-charged FEP, plain hydrogel & charged FEP and pyramidal hydrogel & charged FEP; (**b**) Output voltages and currents of EHTS at different load resistances by shaker tapping; (**c**) The optimization of output power with different load resistances; Output performance of EHTS at −10 °C (**d**), 40 °C (**e**); (**f**) Long-time durability test over 15,000 cycles; (**g**) Output signals of EHTS when be bent for 30°, 60° and 90°; (**h**) Dynamic output of the EHTS attached to a mask when man blowing; (**i**) Output performances of EHTS attached into a mask to detect the dynamic breath of a person.

To explore the influence of the external temperature variations on the output performance of the EHTS, comparative experiments have been designed to test the voltages of the EHTS under −10 °C and 40 °C. The results are shown in Figure 7d,e. With a simple finger compressing stimulation, EHTS can generate the output signals around a peak voltage of 10 V with no noticeable variations among a wide temperature range from −10 °C to 40 °C. The excellent output stabilities under the extreme temperature environment can be mainly attributed to the extraordinary anti-freezing property of the flexible SRT hydrogel, which will further broaden the applicability of the self-powered wearable electronics. In addition, a long-time durability test is also investigated, and EHTS has been tested using a mechanical shaker at a constant frequency of 10 Hz continuously. Hence, EHTS shows excellent durability, and the output voltage remains unchanged after 15,000 cycles. No apparent structural damage occurred during this period. The fabricated EHTS demonstrates great flexibility as well as considerable output performance during bending. As is demonstrated in Figure 7g, EHTS can generate peak voltages around 2.3, 2.9, and 5 V, respectively, when being bent at 30°, 60°, and 90°.

Finally, the fabricated EHTS is assembled into the two filter layers of a mask to monitor dynamic breaths. Although the stimulation strength of man's breath is too weak to activate traditional TENGs, a clear signal with the peak voltage around 2 V can still be acquired by integrating the EHTS between the two filter layers of a mask (Figure 7h). EHTS is further attached in a mask to monitor dynamic breath frequency. Figure 7i significantly demonstrates different peak voltages and frequencies when the tester breaths slow (1 Hz) and fast (~2.5 Hz), which offers a considerable method for healthcare monitoring during the COVID-19 pandemic.

5. Conclusions

In this paper, an electret/hydrogel-based tactile sensor combined with the pyramidal patterned DN SRT ionic hydrogel membrane and FEP electret thin film is proposed and researched in detail. To enhance the temperature tolerance of the DN ionic hydrogel, a LiBr solvent replacement treatment is introduced to fabricate the SRT hydrogel. Due to this improvement method's low freezing point and strong hydration effect, SRT hydrogel demonstrates great anti-freezing and anti-drying properties (from −10 to 40 °C), which are both critical capabilities for wearable electronics and devices. The mechanical stretchability and electrical stability of the SRT hydrogel with double networked structures are further investigated, while ultra-elongation (>1100%) of the hydrogel sample is obtained along with stable resistance responses while stretching. Multi-layered encapsulation and the spin-coating method were utilized to produce the hydrogel membrane and the whole sensors, ensuring the flexibility (including twisting, folding, and rolling for hydrogel membrane), ultra-thin thickness (<1 mm), and excellent transmittance (around 80% in the visible spectral range) of the EHTS. Moreover, the microstructure patterned SRT hydrogel is fabricated with a typical MEMS process, significantly enhancing the triboelectric output performance by 87.5%. Not only is the triboelectrification strengthened, but the electrostatic induction is also boosted. With a customized corona discharging platform, the FEP film is implanted with the negative charges, which increases EHTS output performance efficiency by 77%. The output power of 4.3 μW is obtained by the EHTS with the small size of only 10 mm × 10 mm × 0.9 mm, and the wide temperature tolerance (−10 to 40 °C) of EHTS is achieved simultaneously.

In general, the hydrogel utilized in this research demonstrates great flexibility, outstanding transparency, excellent stretchability, and anti-freezing/drying properties, which provides an ideal electrode and triboelectric friction layer solution for self-powered wearable electronics. Combined with the triboelectrification and electrostatic induction effects, the self-powered tactile sensor proposed in this paper demonstrates larger output performance and considerable sensitivity, and can even be integrated into a mask to monitor the breath during the COVID-19 pandemic. Therefore, EHTS shows great potential in the field of healthcare devices and wearable electronics.

Author Contributions: Conceptualization and methodology, Z.C. (Zhensheng Chen) and J.Y.; validation, J.W. and K.T.; investigation, Y.L.; data curation, H.Z.; writing—review and editing, Z.C. (Zhao Chen), J.W. and K.T.; funding acquisition, J.W. and K.T. All authors have read and agreed to the published version of the manuscript.

Funding: This work is supported by the National Natural Science Foundation of China Grant (Nos. 51705429 & 61801525), the Fundamental. Research Funds for the Central Universities, Guangdong Natural Science Funds Grant (2018A030313400), Space Science and Technology Foundation, Shenzhen Research Plan (JCYJ20180306171637410), 111 Project No. B13044.

Data Availability Statement: The data that support the findings of this study are available from the first author upon reasonable request.

Conflicts of Interest: The authors declare no conflict of interest.

References

1. Gong, S.; Zhang, J.; Wang, C.; Ren, K.; Wang, Z.L. A Monocharged Electret Nanogenerator-Based Self-Powered Device for Pressure and Tactile Sensor Applications. *Adv. Funct. Mater.* **2019**, *29*, 1807618. [CrossRef]
2. Tao, K.; Chen, Z.; Yi, H.; Zhang, R.; Shen, Q.; Wu, J.; Tang, L.; Fan, K.; Fu, Y.; Miao, J.; et al. Hierarchical Honeycomb-Structured Electret/Triboelectric Nanogenerator for Biomechanical and Morphing Wing Energy Harvesting. *Nanomicro Lett.* **2021**, *13*, 123. [CrossRef]
3. Wu, Z.; Ding, H.; Tao, K.; Wei, Y.; Gui, X.; Shi, W.; Xie, X.; Wu, J. Ultrasensitive, Stretchable, and Fast-Response Temperature Sensors Based on Hydrogel Films for Wearable Applications. *ACS Appl. Mater. Interfaces* **2021**, *13*, 21854–21864. [CrossRef]
4. Fan, F.R.; Tian, Z.Q.; Wang, Z.L. Flexible triboelectric generator! *Nano Energy* **2012**, *1*, 328–334. [CrossRef]
5. Kim, H.G.; Hajra, S.; Oh, D.; Kim, N.; Kim, H.J. Additive manufacturing of high-performance carbon-composites: An integrated multi-axis pressure and temperature monitoring sensor. *Compos. Part B Eng.* **2021**, *222*, 109079. [CrossRef]
6. Sahu, M.; Hajra, S.; Kim, H.G.; Rubahn, H.G.; Mishra, Y.K.; Kim, H.J. Additive manufacturing-based recycling of laboratory waste into energy harvesting device for self-powered applications. *Nano Energy* **2021**, *88*, 106255. [CrossRef]
7. Wu, Z.X.; Shi, W.X.; Ding, H.J.; Zhong, B.Z.; Huang, W.X.; Zhou, Y.B.; Gui, X.C.; Xie, X.; Wu, J. Ultrastable, stretchable, highly conductive and transparent hydrogels enabled by salt-percolation for high-performance temperature and strain sensing. *J. Mater. Chem. C* **2021**, *9*, 13668–13679. [CrossRef]
8. Tao, K.; Yi, H.; Yang, Y.; Chang, H.; Wu, J.; Tang, L.; Yang, Z.; Wang, N.; Hu, L.; Fu, Y.; et al. Origami-inspired electret-based triboelectric generator for biomechanical and ocean wave energy harvesting. *Nano Energy* **2020**, *67*, 104197. [CrossRef]
9. Bai, P.; Zhu, G.; Lin, Z.H.; Jing, Q.; Chen, J.; Zhang, G.; Ma, J.; Wang, Z.L. Integrated multilayered triboelectric nanogenerator for harvesting biomechanical energy from human motions. *ACS Nano* **2013**, *7*, 3713–3719. [CrossRef]
10. Guo, H.; Yeh, M.H.; Zi, Y.; Wen, Z.; Chen, J.; Liu, G.; Hu, C.; Wang, Z.L. Ultralight Cut-Paper-Based Self-Charging Power Unit for Self-Powered Portable Electronic and Medical Systems. *ACS Nano* **2017**, *11*, 4475–4482. [CrossRef]
11. Parida, K.; Thangavel, G.; Cai, G.; Zhou, X.; Park, S.; Xiong, J.; Lee, P.S. Extremely stretchable and self-healing conductor based on thermoplastic elastomer for all-three-dimensional printed triboelectric nanogenerator. *Nat. Commun.* **2019**, *10*, 2158. [CrossRef] [PubMed]
12. Li, Z.; Zheng, Q.; Wang, Z.L.; Li, Z. Nanogenerator-Based Self-Powered Sensors for Wearable and Implantable Electronics. *Research* **2020**, *2020*, 8710686. [CrossRef]
13. Meng, B.; Tang, W.; Too, Z.H.; Zhang, X.S.; Han, M.D.; Liu, W.; Zhang, H.X. A transparent single-friction-surface triboelectric generator and self-powered touch sensor. *Energy Environ. Sci.* **2013**, *6*, 3235–3240. [CrossRef]
14. Diteesawat, R.S.; Helps, T.; Taghavi, M.; Rossiter, J. Electro-pneumatic pumps for soft robotics. *Sci. Robot.* **2021**, *6*, eabc3721. [CrossRef]
15. Ma, X.; Zhang, X.; Fang, P. Flexible film-transducers based on polypropylene piezoelectrets: Fabrication, properties, and applications in wearable devices. *Sens. Actuators A Phys.* **2017**, *256*, 35–42. [CrossRef]
16. Jin, T.; Sun, Z.; Li, L.; Zhang, Q.; Zhu, M.; Zhang, Z.; Yuan, G.; Chen, T.; Tian, Y.; Hou, X.; et al. Triboelectric nanogenerator sensors for soft robotics aiming at digital twin applications. *Nat. Commun.* **2020**, *11*, 5381. [CrossRef]
17. Li, G.; Liu, S.; Wang, L.; Zhu, R. Skin-inspired quadruple tactile sensors integrated on a robot hand enable object recognition. *Sci. Robot.* **2020**, *5*, eabc8134. [CrossRef] [PubMed]
18. Wen, D.-L.; Liu, X.; Deng, H.-T.; Sun, D.-H.; Qian, H.-Y.; Brugger, J.; Zhang, X.-S. Printed silk-fibroin-based triboelectric nanogenerators for multi-functional wearable sensing. *Nano Energy* **2019**, *66*, 104123. [CrossRef]
19. He, Z.; Yuan, W. Adhesive, Stretchable, and Transparent Organohydrogels for Antifreezing, Antidrying, and Sensitive Ionic Skins. *ACS Appl. Mater. Interfaces* **2021**, *13*, 1474–1485. [CrossRef]
20. Jiang, Y.; Dong, K.; Li, X.; An, J.; Wu, D.; Peng, X.; Yi, J.; Ning, C.; Cheng, R.; Yu, P.; et al. Stretchable, Washable, and Ultrathin Triboelectric Nanogenerators as Skin-Like Highly Sensitive Self-Powered Haptic Sensors. *Adv. Funct. Mater.* **2020**, *31*, 2005584. [CrossRef]
21. Zhao, X.; Pei, D.; Yang, Y.; Xu, K.; Yu, J.; Zhang, Y.; Zhang, Q.; He, G.; Zhang, Y.; Li, A.; et al. Green Tea Derivative Driven Smart Hydrogels with Desired Functions for Chronic Diabetic Wound Treatment. *Adv. Funct. Mater.* **2021**, *31*, 2009442. [CrossRef]
22. Wu, J.; Wu, Z.; Huang, W.; Yang, X.; Liang, Y.; Tao, K.; Yang, B.R.; Shi, W.; Xie, X. Stretchable, Stable, and Room-Temperature Gas Sensors Based on Self-Healing and Transparent Organohydrogels. *ACS Appl. Mater. Interfaces* **2020**, *12*, 52070–52081. [CrossRef] [PubMed]
23. Taghavi, M.; Stinchcombe, A.; Greenman, J.; Mattoli, V.; Beccai, L.; Mazzolai, B.; Melhuish, C.; Ieropoulos, I.A. Self sufficient wireless transmitter powered by foot-pumped urine operating wearable MFC. *Bioinspir. Biomim.* **2015**, *11*, 016001. [CrossRef]
24. Taghavi, M.; Beccai, L. A contact-key triboelectric nanogenerator: Theoretical and experimental study on motion speed influence. *Nano Energy* **2015**, *18*, 283–292. [CrossRef]
25. Taghavi, M.; Helps, T.; Rossiter, J. Electro-ribbon actuators and electro-origami robots. *Sci. Robot.* **2018**, *3*, eaau9795. [CrossRef]
26. Wu, Z.; Rong, L.; Yang, J.; Wei, Y.; Tao, K.; Zhou, Y.; Yang, B.R.; Xie, X.; Wu, J. Ion-Conductive Hydrogel-Based Stretchable, Self-Healing, and Transparent NO2 Sensor with High Sensitivity and Selectivity at Room Temperature. *Small* **2021**, e2104997. [CrossRef]
27. Zou, Y.; Libanori, A.; Xu, J.; Nashalian, A.; Chen, J. Triboelectric Nanogenerator Enabled Smart Shoes for Wearable Electricity Generation. *Research* **2020**, *2020*, 158953. [CrossRef] [PubMed]

28. Tao, K.; Yi, H.P.; Yang, Y.; Tang, L.H.; Yang, Z.S.; Wu, J.; Chang, H.L.; Yuan, W.Z. Miura-origami-inspired electret/triboelectric power generator for wearable energy harvesting with water-proof capability. *Microsyst. Nanoeng.* **2020**, *6*, 56. [CrossRef]
29. Tao, K.; Zhao, Z.; Yang, Y.; Wu, J.; Li, Y.; Fan, K.; Fu, Y.; Chang, H.; Yuan, W. Development of bipolar-charged electret rotatory power generator and application in self-powered intelligent thrust bearing. *Nano Energy* **2021**, *90*, 106491. [CrossRef]
30. Lai, Y.C.; Hsiao, Y.C.; Wu, H.M.; Wang, Z.L. Waterproof fabric-based multifunctional triboelectric nanogenerator for universally harvesting energy from raindrops, wind, and human motions and as self-powered sensors. *Adv. Sci.* **2019**, *6*, 1801883. [CrossRef]
31. Bi, M.; Wang, S.; Wang, X.; Ye, X. Freestanding-electret rotary generator at an average conversion efficiency of 56%: Theoretical and experimental studies. *Nano Energy* **2017**, *41*, 434–442. [CrossRef]
32. Zhang, L.M.; Han, C.B.; Jiang, T.; Zhou, T.; Li, X.H.; Zhang, C.; Wang, Z.L. Multilayer wavy-structured robust triboelectric nanogenerator for harvesting water wave energy. *Nano Energy* **2016**, *22*, 87–94. [CrossRef]
33. Zhang, X.-S.; Han, M.; Kim, B.; Bao, J.-F.; Brugger, J.; Zhang, H. All-in-one self-powered flexible microsystems based on triboelectric nanogenerators. *Nano Energy* **2018**, *47*, 410–426. [CrossRef]
34. Niu, S.; Wang, X.; Yi, F.; Zhou, Y.S.; Wang, Z.L. A universal self-charging system driven by random biomechanical energy for sustainable operation of mobile electronics. *Nat. Commun.* **2015**, *6*, 8975. [CrossRef]
35. Tao, K.; Tang, L.H.; Wu, J.; Lye, S.W.; Chang, H.L.; Miao, J.M. Investigation of Multimodal Electret-Based MEMS Energy Harvester With Impact-Induced Nonlinearity. *J. Microelectromech. Syst.* **2018**, *27*, 276–288. [CrossRef]
36. Zhou, Y.; Deng, W.; Xu, J.; Chen, J. Engineering Materials at the Nanoscale for Triboelectric Nanogenerators. *Cell Rep. Phys. Sci.* **2020**, *1*, 100142. [CrossRef]
37. Liu, J.; Gu, L.; Cui, N.; Xu, Q.; Qin, Y.; Yang, R. Fabric-Based Triboelectric Nanogenerators. *Research* **2019**, *2019*, 1091632. [CrossRef]
38. Wang, P.; Pan, L.; Wang, J.; Xu, M.; Dai, G.; Zou, H.; Dong, K.; Wang, Z.L. An Ultra-Low-Friction Triboelectric–Electromagnetic Hybrid Nanogenerator for Rotation Energy Harvesting and Self-Powered Wind Speed Sensor. *ACS Nano* **2018**, *12*, 9433–9440. [CrossRef]
39. Wang, X.; Niu, S.; Yi, F.; Yin, Y.; Hao, C.; Dai, K.; Zhang, Y.; You, Z.; Wang, Z.L. Harvesting ambient vibration energy over a wide frequency range for self-powered electronics. *ACS Nano* **2017**, *11*, 1728–1735. [CrossRef]
40. Luo, A.; Zhang, Y.; Dai, X.; Wang, Y.; Xu, W.; Lu, Y.; Wang, M.; Fan, K.; Wang, F. An inertial rotary energy harvester for vibrations at ultra-low frequency with high energy conversion efficiency. *Appl. Energy* **2020**, *279*, 115762. [CrossRef]
41. Zhang, X.; Shan, X.; Xie, T.; Miao, J. A new sensor inspired by the lateral-line system of fish using the self-powered d33 mode piezoelectric diaphragm for hydrodynamic sensing. *Mech. Syst. Signal Process.* **2020**, *141*, 106476. [CrossRef]
42. Wang, Z.L. Nanogenerators, self-powered systems, blue energy, piezotronics and piezo-phototronics—A recall on the original thoughts for coining these fields. *Nano Energy* **2018**, *54*, 477–483. [CrossRef]
43. Gao, L.; Hu, D.; Qi, M.; Gong, J.; Zhou, H.; Chen, X.; Chen, J.; Cai, J.; Wu, L.; Hu, N. A double-helix-structured triboelectric nanogenerator enhanced with positive charge traps for self-powered temperature sensing and smart-home control systems. *Nanoscale* **2018**, *10*, 19781–19790. [CrossRef] [PubMed]
44. Zou, H.; Zhang, Y.; Guo, L.; Wang, P.; He, X.; Dai, G.; Zheng, H.; Chen, C.; Wang, A.C.; Xu, C.; et al. Quantifying the triboelectric series. *Nat. Commun.* **2019**, *10*, 1427. [CrossRef] [PubMed]
45. Zhang, D.; Wang, D.; Xu, Z.; Zhang, X.; Yang, Y.; Guo, J.; Zhang, B.; Zhao, W. Diversiform sensors and sensing systems driven by triboelectric and piezoelectric nanogenerators. *Coord. Chem. Rev.* **2021**, *427*, 213597. [CrossRef]
46. Wang, Y.; Zhang, L.; Lu, A. Highly stretchable, transparent cellulose/PVA composite hydrogel for multiple sensing and triboelectric nanogenerators. *J. Mater. Chem. A* **2020**, *8*, 13935–13941. [CrossRef]
47. Shan, X.; Li, H.; Yang, Y.; Feng, J.; Wang, Y.; Xie, T. Enhancing the performance of an underwater piezoelectric energy harvester based on flow-induced vibration. *Energy* **2019**, *172*, 134–140. [CrossRef]
48. Wang, Z.L.; Chen, J.; Lin, L. Progress in triboelectric nanogenerators as a new energy technology and self-powered sensors. *Energy Environ. Sci.* **2015**, *8*, 2250–2282. [CrossRef]
49. Zhao, L.-C.; Zou, H.-X.; Yan, G.; Liu, F.-R.; Tan, T.; Wei, K.-X.; Zhang, W.-M. Magnetic coupling and flextensional amplification mechanisms for high-robustness ambient wind energy harvesting. *Energy Convers. Manag.* **2019**, *201*, 112166. [CrossRef]
50. Yang, Y.; Han, J.; Huang, J.; Sun, J.; Wang, Z.L.; Seo, S.; Sun, Q. Stretchable Energy-Harvesting Tactile Interactive Interface with Liquid-Metal-Nanoparticle-Based Electrodes. *Adv. Funct. Mater.* **2020**, *30*, 1909652. [CrossRef]
51. Matsuzaki, R.; Tabayashi, K. Highly Stretchable, Global, and Distributed Local Strain Sensing Line Using GaInSn Electrodes for Wearable Electronics. *Adv. Funct. Mater.* **2015**, *25*, 3806–3813. [CrossRef]
52. Chossat, J.B.; Park, Y.L.; Wood, R.J.; Duchaine, V. A Soft Strain Sensor Based on Ionic and Metal Liquids. *IEEE Sens. J.* **2013**, *13*, 3405–3414. [CrossRef]
53. Wang, S.; Ding, L.; Fan, X.W.; Jiang, W.Q.; Gong, X.L. A liquid metal-based triboelectric nanogenerator as stretchable electronics for safeguarding and self-powered mechanosensing. *Nano Energy* **2018**, *53*, 863–870. [CrossRef]
54. Lou, D.; Wang, C.; He, Z.; Sun, X.; Luo, J.; Li, J. Robust organohydrogel with flexibility and conductivity across the freezing and boiling temperatures of water. *Chem. Commun.* **2019**, *55*, 8422–8425. [CrossRef]
55. Sun, H.; Zhao, Y.; Jiao, S.; Wang, C.; Jia, Y.; Dai, K.; Zheng, G.; Liu, C.; Wan, P.; Shen, C. Environment Tolerant Conductive Nanocomposite Organohydrogels as Flexible Strain Sensors and Power Sources for Sustainable Electronics. *Adv. Funct. Mater.* **2021**, *31*, 2101696. [CrossRef]

56. Bai, Y.; Chen, B.; Xiang, F.; Zhou, J.; Wang, H.; Suo, Z. Transparent hydrogel with enhanced water retention capacity by introducing highly hydratable salt. *Appl. Phys. Lett.* **2014**, *105*, 151903. [CrossRef]
57. Ota, H.; Chen, K.; Lin, Y.; Kiriya, D.; Shiraki, H.; Yu, Z.; Ha, T.J.; Javey, A. Highly deformable liquid-state heterojunction sensors. *Nat. Commun.* **2014**, *5*, 5032. [CrossRef] [PubMed]
58. Yun, G.; Tang, S.Y.; Sun, S.; Yuan, D.; Zhao, Q.; Deng, L.; Yan, S.; Du, H.; Dickey, M.D.; Li, W. Liquid metal-filled magnetorheological elastomer with positive piezoconductivity. *Nat. Commun.* **2019**, *10*, 1300. [CrossRef] [PubMed]
59. Russo, S.; Ranzani, T.; Liu, H.; Nefti-Meziani, S.; Althoefer, K.; Menciassi, A. Soft and Stretchable Sensor Using Biocompatible Electrodes and Liquid for Medical Applications. *Soft Robot.* **2015**, *2*, 146–154. [CrossRef] [PubMed]
60. Zhang, Y.S.; Khademhosseini, A. Advances in engineering hydrogels. *Science* **2017**, *356*, eaaf3627. [CrossRef]

micromachines

MDPI

Article

Vibration Energy Harvester Based on Torsionally Oscillating Magnet

Xinyi Wang [1], Jiaxing Li [1], Chenyuan Zhou [1], Kai Tao [2], Dayong Qiao [2] and Yunjia Li [1,*]

[1] School of Electrical Engineering, Xi'an Jiaotong University, Xi'an 710049, China; xywang@stu.xjtu.edu.cn (X.W.); twinklestar2@stu.xjtu.edu.cn (J.L.); ZhouChenyuan@stu.xjtu.edu.cn (C.Z.)

[2] Micro and Nano Electromechanical Systems Laboratory, Northwestern Polytechnical University, Xi'an 710072, China; taokai@nwpu.edu.cn (K.T.); dyqiao@nwpu.edu.cn (D.Q.)

* Correspondence: liyunjia@xjtu.edu.cn

Abstract: Most of the miniaturized electromagnetic vibrational energy harvesters (EVEHs) are based on oscillating proof mass suspended by several springs or a cantilever structure. Such structural feature limits the miniaturization of the device's footprint. This paper presents an EVEH device based on a torsional vibrating magnet over a stack of flexible planar coils. The torsional movement of the magnet is enabled by microfabricated silicon torsional springs, which effectively reduce the footprint of the device. With a size of 1 cm × 1 cm × 1.08 cm, the proposed EVEH is capable of generating an open-circuit peak-to-peak voltage of 169 mV and a power of 6.9 μW, under a sinusoidal excitation of ±0.5 g (g = 9.8 m/s^2) and frequency of 96 Hz. At elevated acceleration levels, the maximum peak-to-peak output voltage is 222 mV under the acceleration of 7 g (±3.5 g).

Keywords: electromagnetic; energy harvester; MEMS; planar coil

Citation: Wang, X.; Li, J.; Zhou, C.; Tao, K.; Qiao, D.; Li, Y. Vibration Energy Harvester Based on Torsionally Oscillating Magnet. *Micromachines* **2021**, *12*, 1545. https://doi.org/10.3390/mi12121545

Academic Editor: Nam-Trung Nguyen

Received: 22 November 2021
Accepted: 9 December 2021
Published: 12 December 2021

1. Introduction

Vibration energy harvesters (VEHs) are devices that transduce vibration kinetic energy to electric power [1]. They have attracted much attention lately due to their potential as alternative power sources to batteries in vibration environments [2]. The vibration energy harvesters can be implemented with different operation principles, such as piezoelectric [3], electromagnetic [4], electrostatic [5], and triboelectric [6]. Among them, electromagnetic vibration energy harvesters (EVEHs) have the advantage of easy design and fabrication, relatively high output voltage and power, as well as good reliability [7]. The transduction of the EVEH is often realized by vibration-induced movement between a permanent magnet and a set of solenoid coils.

Traditional EVEHs can be implemented by fixing a macroscopic magnet or magnet array to the vicinity of a coil, which is a simple and robust approach [8]. However, the macroscopic EVEH is too bulky to power most of the smart electronic devices, giving rise to the strong demand to miniaturize the EVEHs. The typical methods to miniaturize the EVEHs are based on printed circuit board (PCB) [9–11] and microelectromechanical system (MEMS) technologies [12–15], where the coils are often fabricated with deposited planar coils. The main bottleneck for the planar coils is the limited number of turns, as the wires are confined within a single plane [16]. To cope with this limitation, the microfabricated vibrating proof mass is sometimes used with wound coils [17,18] However, such an approach has two issues: Firstly, the assembly process of the microcomponents is difficult; secondly, when the size of the coil is big, many of the turns in the coils cannot effectively cut the magnetic field generated with the micro magnet. In order to deal with these two issues, the authors have proposed a method based on stacked flexible coils, where high-density coils can be formed within the thickness range of 1–2 mm [19]. The low-cost repeating layers of flexible coils also simplifies the assembly and fabrication processes. It has been shown that with flexible coils, the EVEH device was capable of generating an output voltage of 1.56 V, with a volume of 7.5 cm^3.

Toward the goal of miniaturization, the fabrication of the vibrating proof mass has been shifted from PCB to MEMS technologies in a previous study [20]. In these designs, the permanent magnet is suspended by four folded springs, which is a classic type of layout for reducing the resonant frequency of MEMS structures (in addition to cantilever structures). However, it is discovered that these design strategies limit the miniaturization of the device. Therefore, a structure with a torsionally oscillating magnet is proposed in 2021, which has been proven to be as effective as the folded spring design in terms of output power, but with a relatively small footprint [21]. In this work, we will extend our conference contributions in [21] and systematically present the design, implementation, and results of a torsional EVEH device.

2. System Design

2.1. Structure of the EVEH

The operation of the proposed EVEH device is based on the torsional movement of a disc magnet suspended by microfabricated torsional silicon springs, over a stack of high-density flexible coils, as shown in Figure 1a. Figure 1b shows the exploded view of the proposed EVEH device. A disc magnet is glued to the bottom surface of a silicon plate suspended by straight torsional springs. The magnetization direction of the disc magnet is along the z axis, as depicted by the coordinate in Figure 1a. Two types of torsional springs are presented in this work, i.e., the EVEH denoted as T1 consists of a straight torsional spring connecting to the magnet plate via a wide and short silicon beam (spring–plate connector), and device T2 consists of a folded torsional spring consists of three segments of straight springs, as shown in Figure 1c. For the fabrication of the silicon plate and torsional spring, a 200 μm thick monocrystalline silicon wafer is glued to a carrier wafer and etched through using inductively coupled plasma (ICP) reactive ion etching (RIE) process. A flexible coil stack is mounted below the disc magnet, and the coil–magnet distance is precisely controlled by the spacers between the coil and magnet (0.4 mm in this study). Each flexible coil layer consists of two oppositely wound spiral copper wires electroplated on the two sides of a 180 μm thick polyimide film, as shown in Figure 1d. The stack of flexible coils is aligned by the bolt holes and clamped together between two rigid FR4 layers, in order to be connected in series and form a coil with a large number of turns. Four M1 bolts are used to fix all the components. When the EVEH experiences external vibrations, the magnet oscillates torsionally (around the y axis in Figure 1) and generates electrical power in the coil. The torsional movement is mainly asserted on the long torsional beam, and the spring–plate connector will not deform due to its high stiffness. The main innovation of this technology is the torsional movement of the magnet and the formation of a high-density coil. The torsional movement of the magnet is capable of generating a large magnetic field gradient in the coils, with a reduced footprint. The overall footprint of the EVEH device is the same as the device reported in [20], but the functional area of the silicon layer (plate and spring area) is significantly reduced by using the torsional spring design. In this design, a much larger margin area is included in the device layer, which greatly facilitated the assembly process of the device. The stacked coils are formed by simple and low-cost repeating units of flexible coils. With proper technology development, the stacked coils may have the potential to be integrated into the fabrication of MEMS energy harvesters, enabling batch fabrication of the devices. The main design parameters of the EVEH device are listed in Table 1.

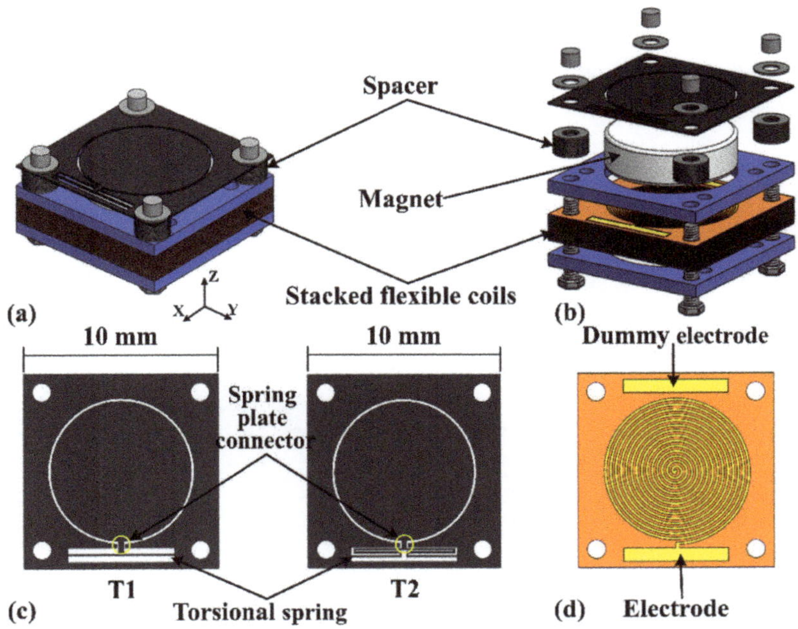

Figure 1. Schematic and exploded illustration of the assembled EVEH. (**a**) schematic illustration of the EVEH device; (**b**) exploded view of the EVEH device; (**c**) two types of torsional spring; (**d**) a layer of flexible coils.

Table 1. Main design parameters of the EVEH device.

	Parameters	Designed	Measured
Overall device	Size of device (L × W × H) (mm)	10 × 10 × 10.7	10 × 10 × 10.8
	Thickness of the wafer (mm)	0.2	0.2
Device T1	Size of torsional spring (L × W) (mm)	4 × 0.08	4 × 0.08
	Size of spring–plate connector (L × W) (mm)	0.4 × 0.8	0.4 × 0.8
	Functional area (mm^2) *	20.3	20.3
Device T2	Size of torsional spring unit (L × W) (mm)	2.04 × 0.08	2.04 × 0.08
	Number of torsional spring folds	2	2
	Size of spring–plate connector (L × W) (mm)	0.5 × 0.4	0.5 × 0.4
	Functional area (mm^2) *	20.6	20.6
Magnet	Diameter (mm)	4.00	4.04
	Thickness (mm)	3.00	3.04
Coil	Numbers of turns	40	40
	Magnet–coil distance (mm)	0.40	0.45

* Functional area indicates the total area occupied by the silicon plate and springs.

2.2. Modeling of the EVEH

According to Newton's second law, the EVEH dynamic behavior can be described using a second-order spring mass damper system as follows:

$$m\frac{\mathrm{d}^2 z(t)}{\mathrm{d}t^2} + c\frac{\mathrm{d}z(t)}{\mathrm{d}t} + kz(t) = m\omega^2 Y \sin(\omega t) \tag{1}$$

$$2\pi f = \sqrt{\frac{k}{m}} \tag{2}$$

where m is the proof mass (magnet) of the EVEH device, c is the total damping coefficient including mechanical and electromagnetic damping, k is the effective stiffness of spring, $z(t)$ is the relative displacement between the magnet and base of the z axis, $y(t) = Y\sin(\omega t)$ is the external vibration applied to the base, and f is the resonant frequency of spring. According to Equation (2), the resonant frequency of the device can be reduced by decreasing the effective stiffness of the spring or increasing the mass of the magnet. For a torsional spring, the effective torsional stiffness can be described as follows [22]:

$$T = \frac{KG\theta}{L} \tag{3}$$

where K is the spring torsion constant dependent on the form and dimensions of the cross section, T is the twisting moment, L is the length of the spring on y direction in Figure 1, G is the shear modulus of the spring, and θ is the torsional angle of spring (twisting around the y direction). According to [22], the value of K is given by

$$K = ab^3\left[\frac{16}{3} - 3.36\frac{b}{a}\left(1 - \frac{b^4}{12a^4}\right)\right] \quad \text{for} \quad a \geq b \tag{4}$$

where a and b are half of the width of the spring and half of the thickness of the spring, respectively. When the magnet moves under external vibration, according to Faraday's law, the induced voltage in a coil within a changing magnetic field is given by

$$V = -N\frac{\mathrm{d}\phi}{\mathrm{d}t} = -NA\frac{\mathrm{d}B}{\mathrm{d}z}\frac{\mathrm{d}z}{\mathrm{d}\theta}\frac{\mathrm{d}\theta}{\mathrm{d}t} \tag{5}$$

where N is the number of turns of the solenoid coil, Φ is the magnetic flux, t is time, A is the area of the coil, and B is the magnetic flux density. Additionally, when the EVEH device is twisting around the y direction (in Figure 1) at a small angle, the value of z equals θ. The magnetic flux density of a disc magnet along the central axis (z axis) can be described as follows [23]:

$$B_z = \frac{B_r}{2}\left\{\frac{d+t}{\sqrt{(d+t)^2 + R^2}} - \frac{d}{\sqrt{d^2 + R^2}}\right\} \tag{6}$$

where B_z is the magnetic flux density of the z axis, B_r is the remnant magnetic flux density, R and t are the radius and thickness of the disc magnet, and d is the distance from the surface of the magnet. For a torsional spring, Equation (6) can be written as

$$B_z = \frac{B_r}{2}\left\{\frac{r\sin\theta + d_0 + t}{\sqrt{(r\sin\theta + d_0 + t)^2 + R^2}} - \frac{r\sin\theta + d_0}{\sqrt{(r\sin\theta + d_0)^2 + R^2}}\right\}\cos\theta \tag{7}$$

where r is the distance between the center of the magnet and torsional spring, and d_0 is the initial distance between magnet and coil. Figure 2 shows the analytical calculated magnetic flux density along the z-axis as a function of θ, and the magnetic field gradient ($\mathrm{d}B/\mathrm{d}\theta$) is linear at a small angle.

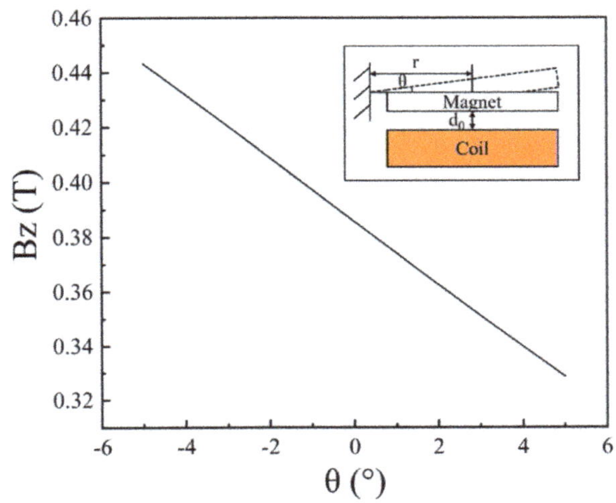

Figure 2. Curve of magnetic flux density along the z axis as a function of θ.

Equation (4) indicates that under a certain external vibration, the following parameters can be increased in order to enhance the induced voltage: the number of coil turns (N), the area of the coil (A), the magnetic field gradient ($dB/d\theta$) around the coil, or the torsional spring design ($d\theta/dt$). However, when designing a miniaturized EVEH device, the coil area (A) and magnetic flux gradient ($dB/d\theta$) are always limited by the device design (size and materials) and fabrication technologies. Therefore, a more effective way to increase induced voltage is by increasing the number of turns in the coil and the vibration amplitude of the EVEH device's movable part. The vibration amplitude can be improved by different spring designs. In comparison, the number of coil turns (based on the planar coil technology) is difficult to improve in most of the reported MEMS EVEH devices.

In order to study the dynamic characteristics of the EVEH device with a torsional spring, a FEM modal analysis of the proposed EVEH was performed with COMSOL Multiphysics software, as shown in Figure 3. In the FEM model, the simplification of the anchoring frame structure was carried out by applying fixed constraints to the end of the springs. This simplification is valid as the torsional spring is an in-plane spring. The material parameters of the silicon used in the FEM model were Young's Modulus of 170 GPa and Poisson's ratio of 0.28. The torsional vibration mode (the main resonance mode during operation, as the device is subjected to a uniform z-direction acceleration, and other resonant modes have a very low possibility be excited with the z-axis vibration) of the EVEHs is shown in Figure 3: T1 device twisting mode around the x-axis at 128.49 Hz, and T2 device twisting mode around the x-axis at 107.97 Hz. However, the FEM simulation of the torsional mode is not accurate, for the mode is influenced by the residual stress of the spring, and the complex stress distribution is difficult to include in the FEM model. The factor also ignored in the FEM model is the slight geometrical asymmetry caused by the device assembly process (e.g., the eccentric position of the magnet caused by the gluing process).

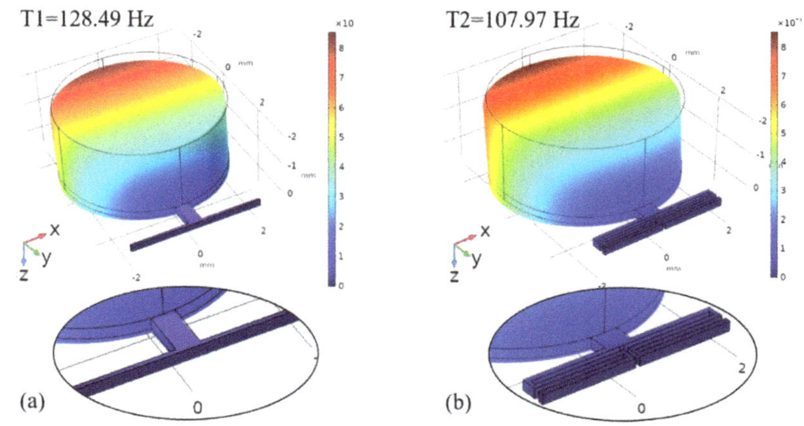

Figure 3. FEM modal analysis of the EVEH: (**a**) T1 device (128.49 Hz); (**b**) T2 device (107.97 Hz).

Figure 4 shows the distribution of the magnetic flux over the coil at different magnet–coil distances: (a) 0.15 mm and (b) 0.75 mm. It was obtained from a 2D axisymmetric FEM model of the simplified magnet–coil structure. In the figure, M and C represent the magnet and coil, respectively, and the arrowed lines indicate the magnetic flux lines of the magnet with z-axis magnetization. When the distance between magnet and coil is increased from 0.15 mm to 0.75 mm, the magnetic flux density along the z axis is decreased from 0.34 T to 0.29 T. This change of the magnetic flux in the coil represents the origin of the induction process of the EVEH under vibration excitation.

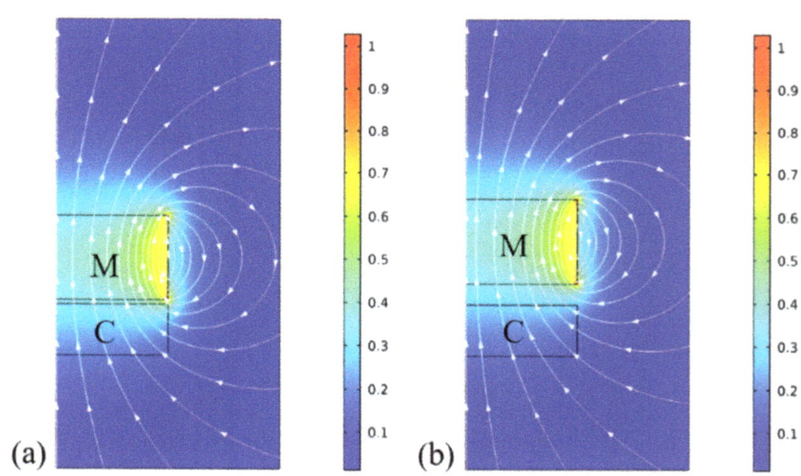

Figure 4. The magnetic flux of the magnet–coil structure based on a 2D axisymmetric FEM model at different magnet–coil distances: (**a**) 0.15 mm; (**b**) 0.75 mm.

3. Experimental

3.1. Assembly Process

Figure 5 shows the assembly process of the proposed EVEH. As shown in Figure 5a, the flexible coil layers were stacked onto a rigid FR4 frame, which contained electrical feedthroughs on the bottom. Subsequently, four bolts were used to fix the stacked coils and bottom package, as shown in Figure 5b. Afterward, the disc magnet was glued to a circular silicon plate connected to the torsional spring, as shown in Figure 5c. The intermediate package and the silicon layer device were then installed with four bolts, as shown in Figure 5d. Finally, the top cover of the package was installed by using four bolts, as shown in Figure 5e. The package of the EVEH device in this work was an anodized aluminum case for both mechanical protection and electrical insulation.

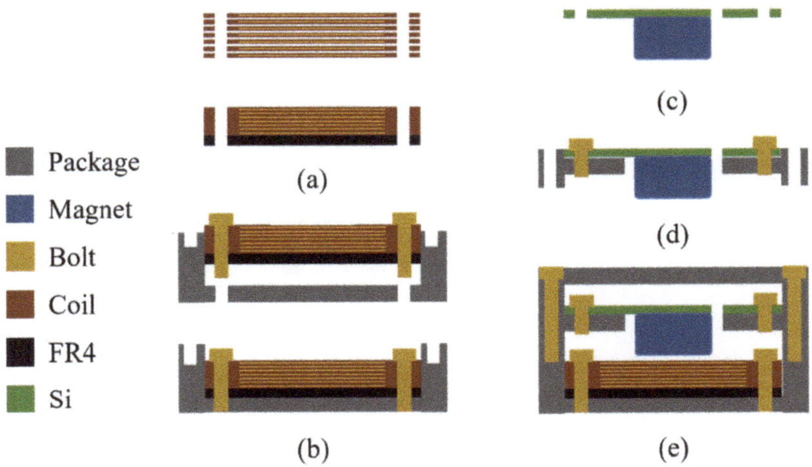

Figure 5. Assembly process of: (**a**) the stacked flexible coils; (**b**) the bottom package; (**c**) the magnet and device layer; (**d**) the intermediate package; (**e**) the package top cover.

3.2. Measurement Setup

Figure 6 shows the measurement setup for characterizing the dynamic behavior of the fabricated EVEH. In the setup, the EVEH device was mounted on a vibration shaker, which was excited by a sinusoidal signal produced by the signal generator and amplified by a power amplifier. The real-time excitation acceleration was monitored by an accelerometer mounted underneath the EVEH by a customized installation fixture. Experimental data including the EVEH output and applied acceleration signals were collected simultaneously by a NI USB-6211 data acquisition board and LabVIEW software, with a sampling rate of 1000 samples per second. A fourth-order Butterworth low pass filter with a cut-off frequency of 500 Hz was used to remove the high-frequency noise, implemented in the LabVIEW software. The acquired voltage in this work has a measurement error of $\pm$0.05 mV, derived from the standard deviation of 594 periods of sinusoidal voltage output.

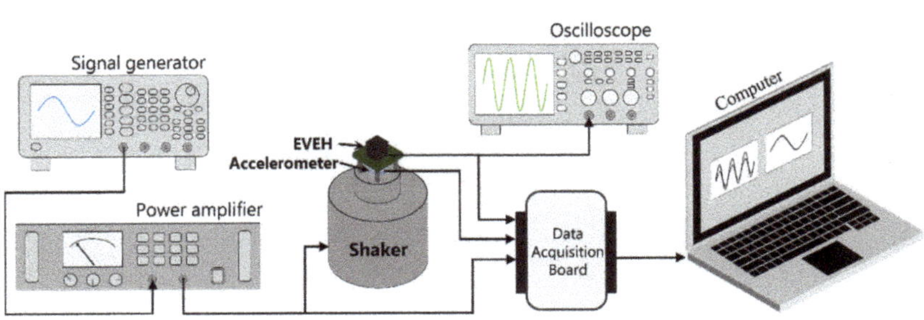

Figure 6. Schematic illustration of the measurement setup.

4. Results and Discussions

4.1. Fabrication Results of the EVEH Device

The designed EVEH device (T1 device) was successfully fabricated and assembled, as shown in Figure 7a. Additionally, for the T2 device, the difference is the spring type shown in Figure 1c. The NdFeB disc magnet was glued to the bottom of the circular silicon microplate as the proof-mass (not visible in the photo). The magnet proof-mass was suspended by a long torsional beam above a stack of flexible coils. Under external vibration excitation, the relative position of the magnet and the coil changed to generate a changing magnetic field, and an electrical voltage was induced in the coil. The stacked flexible coils (clamped between two rigid FR4 frames) were mounted on the bottom of the EVEH device. Additionally, for a large number of turns, the thickness of the stacked flexible coils is reasonable. The thin substrates make it possible to have high-density coils close to the magnet. This is essential for the miniaturized device, as the magnetic field decays exponentially from the outer surface of the magnet. Through this coil design, the EVEH device can achieve high output power with a minimum footprint. The proposed technology is a general technology that can address the limitations of planar coils widely used in MEMS devices. This method is simple and cost effective because the coil stack is composed of repeating low-cost flexible layers. In addition, the distance between the magnet and the coils could be adjusted by the spacers between them. The flexible coil had an identical spiral coil on each side but with opposite winding directions, connected by the via in the center of the coil, as shown in Figure 7b. The electrode and dummy electrode were located at the edge of the film, and the positions were reversed on the other side of the coil layer. When two layers were clamped together (the electrodes were connected), a coil with four windings connected in series was formed. The dummy electrodes were patterned on each side of the coil layer to maintain the mechanical symmetry of the coil layer during stacking and clamping. For comparing the performance of four types of spring design, the coil size and the overall footprint of the EVEH in this work is the same as the device reported in [20], but the functional area of the silicon layer is greatly reduced, which has the potential to the goal of miniaturization.

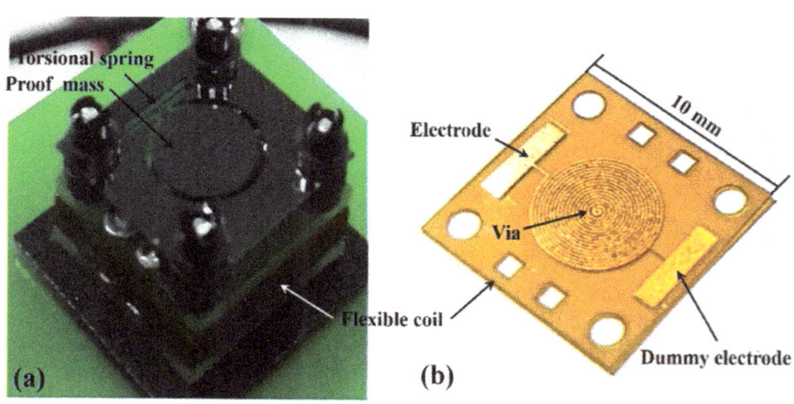

Figure 7. Photos of (**a**) assembled EVEH device (T1) and (**b**) a layer of the flexible coil.

4.2. Open-Circuit Frequency Domain Response of the EVEH Device

Figure 8 shows the open-circuit peak-to-peak output voltage of the assembled EVEH device as a function of frequency, under a sinusoidal excitation (the most common vibration signal existing in both nature and industry), with an amplitude of ± 0.5 g. The output voltage of the EVEH increases gradually as the frequency of the excitation vibration increases toward the resonant frequency, reaches a maximum value at resonance and then decreases again as the frequency further increases. For device T1, at the resonant frequency of 104 Hz, the output voltage reaches the maximum peak-to-peak value of 138.9 mV. For device T2, the maximum open-circuit peak-to-peak output voltage is 169 mV at the resonant frequency of 96 Hz. It is known that most of the ambient vibrations concentrate in the frequency band of 1–200 Hz, the resonant frequency around 100 Hz of the devices T1 and T2 may find applications in several industrial scenarios such as power transformers, transmission lines, and power inductors. In comparison, the open-circuit peak-to-peak output voltage of the EVEH devices presented in [20] is 208.3 mV at the frequency of 143 Hz for D1 and 149.3 mV at the frequency of 156 Hz for D2. Additionally, the functional area of T1, T2, D1, and D2 are 20.3 mm^2, 20.6 mm^2, 23.4 mm^2 and 24.1 mm^2, respectively. Thus, the torsional springs with a reduced functional area of the silicon layer and lower frequency generate a higher output voltage.

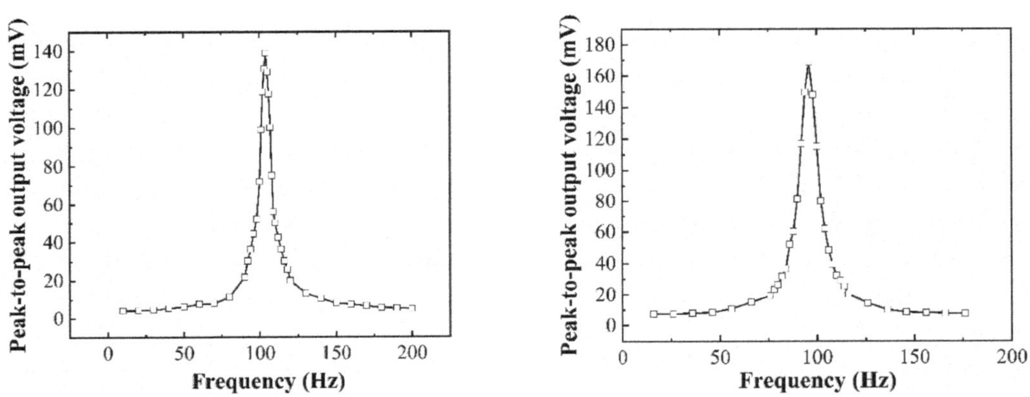

Figure 8. The peak-to-peak open-circuit output voltage of the EVEH as a function of frequency: (**a**) T1 device; (**b**) T2 device.

The presented energy harvester (T1 device) achieves 10 times higher maximum output voltage at 3.8 times lower resonant frequency, compared with a two-degree-of-freedom EVEH based on coils suspended by springs [24], and almost 188 times higher maximum output voltage at 1.2 times lower resonant frequency, compared with an EVEH based on magnet suspended by folded springs reported in 2014 [25]. Compared with the traditional planar coil used in MEMS or PCB technologies, the optimized performance of the EVEH device results from the stacked flexible coils, which greatly enhance the number of turns of the coils.

4.3. Impedance Matching

To study the optimal closed-circuit behavior of the EVEH, matching resistors between 1 Ω and 1000 Ω were connected to the output of the energy harvester. Through the measured closed-circuit voltage, the output power of the EVEH can be described as

$$P = \frac{V_{rms}^2}{R_L} = \frac{V_{pp}^2}{8R_L} \tag{8}$$

where V_{rms} is the root-mean-square voltage over the matching resistor, R_L is the resistance of the matching resistor, and V_{pp} is the peak-to-peak voltage over the matching resistor.

Figure 9 shows the closed-circuit peak-to-peak output voltage and output power of the EVEH device as a function of the load resistance, under a sinusoidal excitation with an amplitude of ±0.5 g and frequency of 104 Hz for T1 device (96 Hz for T2 device). For both types of devices, the internal (coil) resistance of the EVEH under test is 128.7 Ω. With the increase in load resistance, the output voltage of the EVEH first increases sharply and then decreases gradually. For the T1 device, the maximum power measured is 4.6 μW at the load resistance of 128.7 Ω and the resonant frequency of 104 Hz. In comparison, the T2 device has a maximum power measured as 6.9 μW at the load resistance of 128.7 Ω and the resonant frequency of 96 Hz. Additionally, with the same coil resistance, both designs of the proposed torsional EVEHs have lower frequencies than previous work in [20]; the maximum output power is 10.5 μW (143 Hz) for the D1 device and 5.4 μW (156 Hz) for the D2 device.

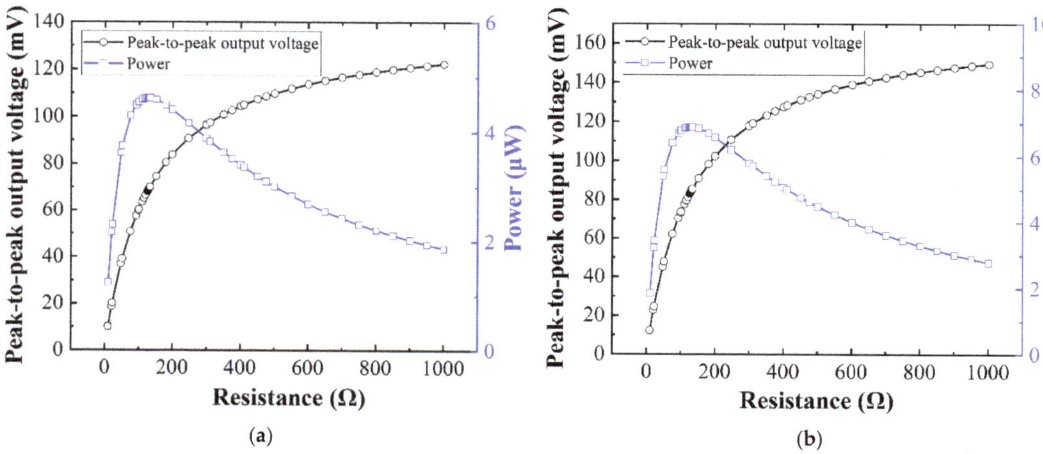

Figure 9. The peak-to-peak output voltage and power of the EVEH as a function of load resistance: (**a**) T1 device; (**b**) T2 device.

4.4. Closed-Circuit Frequency-Domain Response of the EVEH Device

Figure 10 shows the closed-circuit peak-to-peak output voltage and output power of the EVEH device when connected to a load resistance of 128.7 Ω. The measurements were performed under sinusoidal excitation of ±0.5 g. For the T1 device, the maximum peak-to-peak closed-circuit output voltage is 69.4 mV at 104 Hz (resonant frequency). For device T2, the maximum peak-to-peak closed-circuit output voltage of the EVEH is 84 mV at 96 Hz (resonant frequency). Due to the similar values of the coil and load resistance, the closed-circuit output voltage is approximately half of the open-circuit voltage in Figure 8. The maximum power device T1 and T2 are 4.6 μW at the frequency of 104 Hz and 6.9 μW at the frequency of 96 Hz, respectively. The performance of the EVEH device is improved, compared with recently reported work with similar technologies, in terms of output voltage, power, and resonant frequency (low resonant frequency is preferable for matching to the low-frequency ambient vibrations). For example, Zhang et al. reported an EVEH based on MEMS technology with an output power of 0.82 μW at the resonant frequency of 350 Hz under the acceleration of 4.5 g [26]. Tao et al. reported a MEMS electromagnetic vibration energy harvester with two degrees of freedom; the maximum amplitude of the output voltage in the study is 6.5 mV at the resonant frequency of 391 Hz [24]. The maximum closed-circuit output power of 4.6 μW (T1 device) is 4792 times higher than the maximum closed-circuit output power of 0.96 nW with an optimum load resistance of 12 kΩ at 0.12 g vibration reported in [24]. Since the volume of the EVEH device and external excitation vibration (A_0) have a greater impact on device performance, to accurately describe it, the normalized power density (NPD) is given by

$$\text{NPD} = \frac{P}{\text{Volume} \cdot A_0^2} \tag{9}$$

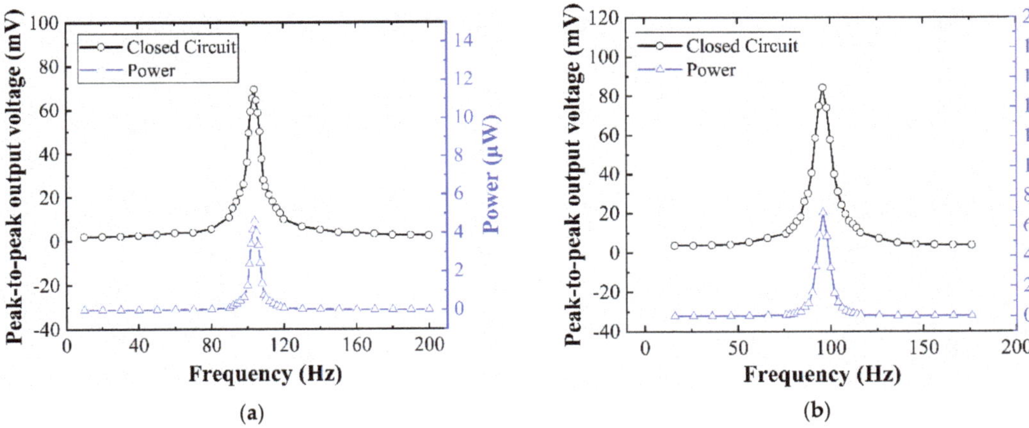

Figure 10. The peak-to-peak closed-circuit output voltage and power of the EVEH device as a function of frequency: (**a**) T1 device; (**b**) T2 device.

The calculated NPD values of the T1 device and T2 device are 17.04 ($\mu W/cm^3/g^2$) and 25.56 ($\mu W/cm^3/g^2$), respectively. Additionally, the NPD value is 38.89 ($\mu W/cm^3/g^2$) for D1 device and 20 ($\mu W/cm^3/g^2$) for D2 device [20]. Tao et al. reported a MEMS electromagnetic vibration energy harvester with two degrees of freedom in 2016 with the NPD value of up to 0.23 ($\mu W/cm^3/g^2$) at the resonant frequency of 391 Hz from 0.02 to 0.2 g [24]. Liu et al. reported an in-plane approximated nonlinear MEMS electromagnetic energy harvester in 2014 with the NPD value of 0.62×10^{-2} ($\mu W/cm^3/g^2$) at the resonant frequency of 146.5 Hz and acceleration of 3 g [25]. Compared with the work reported in [24], the coil resistance in this work is much smaller, so the increase in closed-circuit output power is much bigger than the square of the increase in open-circuit output voltage (see discussions about Figure 8). In addition, with the same footprint, the device T2 has increased its NPD by 50%, compared with T1, indicating the good potential of optimization for the torsional design of the EVEH device. For future optimization of the springs, coils, and magnets, the output power of the proposed EVEH device can be further enhanced.

4.5. Acceleration Test

The dependence of the EVEHs output on the acceleration level was studied, and the open-circuit and closed-circuit peak-to-peak output voltages of the EVEH device were measured as a function of acceleration, as shown in Figure 11. All data were obtained at EVEH's resonant frequency of 104 Hz, for the T1 device, and 96 Hz for the T2 device; the applied peak-to-peak acceleration increases from 1 g to 7 g (step length is 1 g). At acceleration higher than 7 g, the distance between coil and magnet must be increased to reduce the potential collision. The magnet–coil distance can be increased by simply changing the thickness of the spacers. At an acceleration of 7 g, the maximum peak-to-peak open-circuit output voltage is 185.2 mV and 222 mV for device T1 for T2, respectively. Compared with a previous work, also based on microfabricated springs, where the maximum peak-to-peak open-circuit output voltage is 333.1 mV and 225.5 mV at 7 g [20], the output voltage–acceleration relationship in this work shows a similar trend but lower maximum voltage at higher acceleration levels. The output voltage in [20] increases almost linearly as the acceleration increases, whereas the output voltage of devices shows a gradually decreasing rate of enhancement as the acceleration increases. The possible explanation for this phenomenon is that the shear strain induced in the single-beam torsional spring is much larger than the shear strain in the multibeam folded spring, at the same acceleration. Therefore, at elevated acceleration, nonlinearity in the torsional spring arises and limits the deflection angle of the magnet, which reduces the amplitude of enhancement of the output voltage. In addition, it is increasingly difficult for the magnet to maintain its movement perpendicular to the coil plane (z direction) as the vibration excitation acceleration increases, and there will be a tendency to twist in the x or y direction. The torsional vibration tendency will reduce the working efficiency of the device, thereby reducing the output voltage of the device. The closed-circuit output voltage shows a similar linear characteristic to the open-circuit output but with half of the amplitude of the open-circuit output.

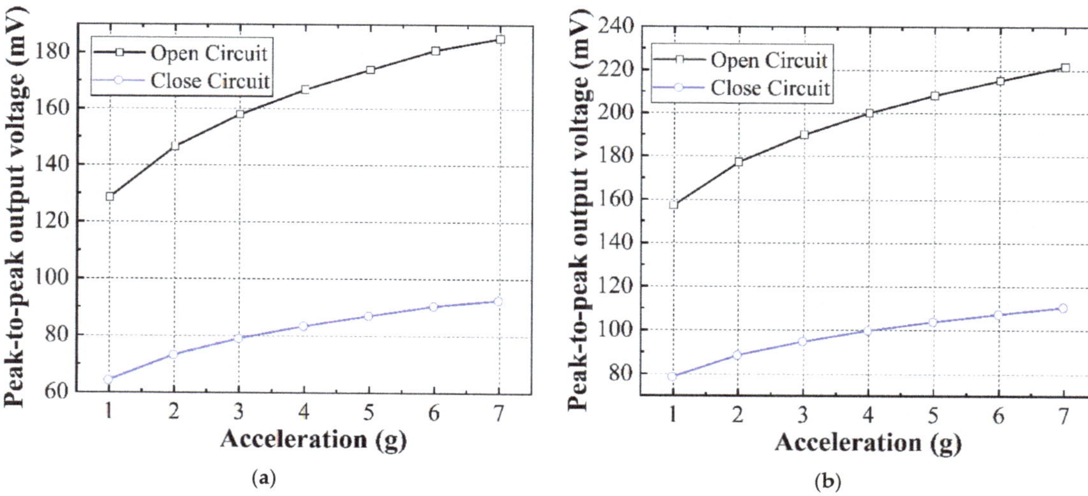

Figure 11. The peak-to-peak output voltage measured in the open circuit and closed circuit at different accelerations: (**a**) T1 device; (**b**) T2 device.

5. Conclusions

In this work, an EVEH device was designed and fabricated based on hybrid PCB and MEMS technology. The proposed EVEH was composed of a stack of flexible coils and a disc permanent magnet mounted vertically. The assembled EVEH was excited by sinusoidal accelerations and its dynamic behavior was characterized. At a sinusoidal acceleration with a peak-to-peak value of 1 g (±0.5 g, ±4.9 m/s^2), a maximum output peak-to-peak voltage of 169 mV and output power of 6.9 μW are realized at a resonant frequency of 96 Hz. At an elevated acceleration of 7 g (±3.5 g), a maximum output peak-to-peak voltage of 222 mV is realized. In the future, the bonding technologies between the stacked coils and silicon proof mass will be explored.

Author Contributions: Conceptualization, K.T. and Y.L.; methodology, X.W. and J.L.; formal analysis, D.Q.; investigation, X.W., J.L. and K.T.; data curation, C.Z.; writing—original draft preparation, X.W.; writing—review and editing, Y.L. All authors have read and agreed to the published version of the manuscript.

Funding: This research received no external funding.

Data Availability Statement: The data presented in this study are available in [insert article or supplementary material here].

Acknowledgments: The authors would like to thank Junyuan Wang for fruitful discussions on modeling the EVEH devices.

Conflicts of Interest: The authors declare no conflict of interest.

References

1. Li, Y.J.; Tao, K.; George, B.; Tan, Z.C. Harvesting Vibration Energy: Technologies and Challenges. *IEEE Ind. Electron. Mag.* **2021**, *15*, 30–39. [CrossRef]
2. Li, M.X.; Deng, H.C.; Zhang, Y.F.; Li, K.X.; Huang, S.J.; Liu, X.W. Ultra-Low Frequency Eccentric Pendulum-Based Electromagnetic Vibrational Energy Harvester. *Micromachines* **2020**, *11*, 1009. [CrossRef]
3. Yang, Z.B.; Zhou, S.X.; Zu, J.; Inman, D. High-Performance Piezoelectric Energy Harvesters and Their Applications. *Joule* **2018**, *2*, 642–697. [CrossRef]
4. Maharjan, P.; Bhatta, T.; Rasel, M.S.; Salauddin, M.; Rahman, M.T.; Park, J.Y. High-performance cycloid inspired wearable electromagnetic energy harvester for scavenging human motion energy. *Appl. Energy* **2019**, *256*, 113987. [CrossRef]
5. Suzuki, Y. Recent Progress in MEMS Electret Generator for Energy Harvesting. *IEEJ Trans. Electr. Electron. Eng.* **2011**, *6*, 101–111. [CrossRef]
6. Wang, Z.L. Triboelectric Nanogenerators as New Energy Technology for Self-Powered Systems and as Active Mechanical and Chemical Sensors. *ACS Nano* **2013**, *7*, 9533–9557. [CrossRef]
7. Tan, Y.S.; Dong, Y.; Wang, X.H. Review of MEMS Electromagnetic Vibration Energy Harvester. *J. Microelectromech. Syst.* **2017**, *26*, 1–16. [CrossRef]
8. Glynne-Jones, P.; Tudor, M.J.; Beeby, S.P.; White, N.M. An electromagnetic, vibration-powered generator for intelligent sensor systems. *Sens. Actuator A-Phys.* **2004**, *110*, 344–349. [CrossRef]
9. Hatipoglu, G.; Urey, H. FR4-based electromagnetic energy harvester for wireless sensor nodes. *Smart Mater. Struct.* **2010**, *19*, 015022. [CrossRef]
10. Lee, B.C.; Rahman, M.A.; Hyun, S.H.; Chung, G.S. Low frequency driven electromagnetic energy harvester for self-powered system. *Smart Mater. Struct.* **2012**, *21*, 125024. [CrossRef]
11. Roundy, S.; Takahashi, E. A planar electromagnetic energy harvesting transducer using a multi-pole magnetic plate. *Sens. Actuator A Phys.* **2013**, *195*, 98–104. [CrossRef]
12. Saadon, S.; Sidek, O. A review of vibration-based MEMS piezoelectric energy harvesters. *Energy Conv. Manag.* **2011**, *52*, 500–504. [CrossRef]
13. Podder, P.; Constantinou, P.; Amann, A.; Roy, S. Frequency adjustable MEMS vibration energy harvester. In Proceedings of the 27th Micromechanics and Microsystems Europe (MME) Workshop, Cork, Ireland, 28–30 August 2016.
14. Shi, Q.F.; Wang, T.; Lee, C. MEMS Based Broadband Piezoelectric Ultrasonic Energy Harvester (PUEH) for Enabling Self-Powered Implantable Biomedical Devices. *Sci. Rep.* **2016**, *6*, 24946. [CrossRef] [PubMed]
15. Toshiyoshi, H.; Ju, S.; Honma, H.; Ji, C.H.; Fujita, H. MEMS vibrational energy harvesters. *Sci. Technol. Adv. Mater.* **2019**, *20*, 124–143. [CrossRef] [PubMed]
16. Wang, P.H.; Tanaka, K.; Sugiyama, S.; Dai, X.H.; Zhao, X.L.; Liu, J.Q. A micro electromagnetic low level vibration energy harvester based on MEMS technology. *Microsyst. Technol.* **2009**, *15*, 941–951. [CrossRef]
17. Mallick, D.; Amann, A.; Roy, S. High Figure of Merit Nonlinear Microelectromagnetic Energy Harvesters for Wideband Applications. *J. Microelectromech. Syst.* **2017**, *26*, 273–282. [CrossRef]
18. Khan, F.U. A vibration-based electromagnetic and piezoelectric hybrid energy harvester. *Int. J. Energy Res.* **2020**, *44*, 6894–6916. [CrossRef]
19. Li, Y.J.; Cao, Q.; Zhang, W.; Zhang, Y.; Cao, J.A. A miniaturized electromagnetic energy harvester with volt-level output based on stacked flexible coils. *Smart Mater. Struct.* **2018**, *27*, 115040. [CrossRef]
20. Li, Y.J.; Li, J.X.; Yang, A.J.; Zhang, Y.; Jiang, B.X.; Qiao, D.Y. Electromagnetic Vibrational Energy Harvester With Microfabricated Springs and Flexible Coils. *IEEE Trans. Ind. Electron.* **2021**, *68*, 2684–2693. [CrossRef]
21. Wang, X.; Li, J.; Zhou, C.; Tao, K.; Qiao, D.; Li, Y. Torsional Electromagnetic Vibrational Energy Harvester Based on Stacked Flexible Coils. In Proceedings of the IECON 2021—47th Annual Conference of the IEEE Industrial Electronics Society, Toronto, ON, Canada, 13–16 October 2021.
22. Young, W.C.; Budynas, R.G. *Roark's Formulas for Stress and Strain*, 7th ed.; McGraw-Hill: New York, NY, USA, 2002.
23. Yang, B.; Lee, C.; Xiang, W.F.; Xie, J.; He, J.H.; Kotlanka, R.K.; Low, S.P.; Feng, H.H. Electromagnetic energy harvesting from vibrations of multiple frequencies. *J. Micromech. Microeng.* **2009**, *19*, 035001. [CrossRef]
24. Tao, K.; Wu, J.; Tang, L.H.; Xia, X.; Lye, S.W.; Miao, J.M.; Hu, X. A novel two-degree-of-freedom MEMS electromagnetic vibration energy harvester. *J. Micromech. Microeng.* **2016**, *26*, 035020. [CrossRef]
25. Liu, H.C.; Qian, Y.; Wang, N.; Lee, C.K. An In-Plane Approximated Nonlinear MEMS Electromagnetic Energy Harvester. *J. Microelectromech. Syst.* **2014**, *23*, 740–749. [CrossRef]
26. Zhang, Q.; Kim, E.S. Microfabricated Electromagnetic Energy Harvesters With Magnet and Coil Arrays Suspended by Silicon Springs. *IEEE Sens. J.* **2016**, *16*, 634–641. [CrossRef]

micromachines

Article

Electromagnetic Vibration Energy Harvester with Tunable Resonance Frequency Based on Stress Modulation of Flexible Springs

Yunjia Li [1,*], Chenyuan Zhou [1], Qi Cao [1], Xinyi Wang [1], Dayong Qiao [2] and Kai Tao [2,*]

1 School of Electrical Engineering, Xi'an Jiaotong University, Xi'an 710049, China; zhouchenyuan@stu.xjtu.edu.cn (C.Z.); caoqisky@126.com (Q.C.); xywang@stu.xjtu.edu.cn (X.W.)
2 Micro and Nano Electromechanical Systems Laboratory, Northwestern Polytechnical University, Xi'an 710072, China; dyqiao@nwpu.edu.cn
* Correspondence: liyunjia@xjtu.edu.cn (Y.L.); taokai@nwpu.edu.cn (K.T.)

Abstract: This paper presents a compact electromagnetic vibrational energy harvester (EVEH) with tunable resonance frequency. The resonance frequency of the EVEH is tuned by adjusting the axial stress in the flexible polymeric springs, which is realized by physically pulling and pushing the springs. The stress tuning functionality is realized with a compact structure with small volume. The total frequency tuning range of the proposed EVEH is 56 Hz (74–130 Hz), which is 64% of the natural resonance frequency of the EVEH (88 Hz). It is found that the tensile stress increases the resonance frequency of the EVEH, while the compressive stress firstly reduces the resonance frequency and then increases the resonance frequency due to buckling.

Keywords: polymer beam; vibration energy harvesting; tunable resonant frequency

Citation: Li, Y.; Zhou, C.; Cao, Q.; Wang, X.; Qiao, D.; Tao, K. Electromagnetic Vibration Energy Harvester with Tunable Resonance Frequency Based on Stress Modulation of Flexible Springs. *Micromachines* **2021**, *12*, 1130. https://doi.org/10.3390/mi12091130

Academic Editor: Hamed Farokhi

Received: 5 September 2021
Accepted: 16 September 2021
Published: 20 September 2021

1. Introduction

Electromagnetic vibration energy harvesters (EVEHs) are devices capable of converting vibration kinetic energy to electrical power [1,2]. The transduction process usually involves relative movement between a permanent magnet and a solenoid coil induced by external vibration. Most of the reported EVEH devices are resonant structures that output maximum power only when its mechanical resonance frequency matches the external vibration frequency. Under off-resonance conditions, the output power of the EVEH decreases significantly. However, the frequency matching is often difficult because the frequency of the environmental vibration sources is often random and unpredictable [3]. Even if the source vibration frequency is constant, it is still difficult to realize the frequency matching due to uncertainties and inaccuracies of the fabrication processes and geometrical parameters of the devices [4].

Two major approaches to improve the frequency matching between the VEH and the vibration sources are the frequency tuning and band broadening techniques. The frequency tuning techniques can be realized by both mechanical tuning and electrical tuning approaches. Mechanical tuning approaches can be implemented by different structures and technologies, but the basic principle is mainly based on tuning the resonance frequency of the VEH by changing certain physical parameters of the device such as the mass or the stiffness [5,6]. Electrical tuning approaches is mainly based on load tuning or frequency tracking technologies. However, Roundy et al. [7] showed that for the active frequency tuning techniques, the energy consumed by the tuning actuators is sometimes more than the energy generated by the VEH, which limits the application of these approaches. The frequency band broadening techniques can be realized by four main different configurations: multi-mode devices, multi-frequency devices, nonlinear devices and bistable devices. The multi-mode device uses different resonance modes of the VEH to harvest energy in different frequency bands [8–12]. Nevertheless, the operation of this type of device is still limited to several rather narrow frequency bands. To extend the number of these bands, the VEH can be implemented using array of structures with different size to realize multi-frequency operation [13–15]. These bandwidth tuning techniques above extend the frequency band of the VEHs to several fixed bands. However, for excitation vibrations with arbitrary frequencies, it is still possible that the VEH's operation band mismatches the excitation frequency. In addition to the multi-mode or multi-frequency methods, the frequency band of the VEH can also be broadened by utilizing nonlinear springs. It has been shown that the cubic stiffness of the springs effectively broadens the bandwidth of the VEH [16,17]. However, one of the deficiencies of the nonlinear VEH is the hysteresis in the frequency response, i.e., different sweep-up and sweep-down characteristics in the frequency domain. In comparison, the bistable broadband energy harvester has a better-defined frequency response to vibrations, but the excitation vibration must be beyond certain threshold to reliably operate this type of device.

From the literature above, it can be concluded that the main goal for the frequency tuning technologies is large tuning range or wide bandwidth. However, the working frequency range of many reported works are still limited to around 10 Hz. In addition, many of the proposed devices are still proof-of-concept prototypes which are rather bulky in size. In this paper, we propose a compact EVEH device with large tunable resonance frequency, based on soft polymer springs and stacked polymer coils. The frequency tuning function of the EVEH is realized by a compact tuning structure that changes the stress within the polymeric springs. The stacking of repeating thin polymer coils enables high output voltage of the EVEH device and the resonance frequency of the EVEH is tuned by modulating the axial stress in the polymer springs. The remainder of the paper is organized as follows: Section 2 reports the design of the EVEH device, Section 3 describes the experimental procedures of the work, Section 4 presents the results and discussions and Section 5 is the conclusion of the work.

2. System Design

2.1. Concept of the Device

The proposed EVEH is schematically illustrated in Figure 1a,b. A disc magnet is suspended by two straight polyimide springs over a stack of flexible coil layers. The details of the flexible coils are detailed in another work [17]. When the EVEH is subjected to external vibration, the magnet oscillates and generates a changing magnetic field around the coil. Consequently, electrical power is generated in the coil via induction. To minimize the footprint of the device, the disc magnet is glued onto a 1 mm-thick reinforcement structure which provides space for the springs to oscillate. The polyimide springs are laser-cut straight beams with a length of 5.5 mm, a width of 1 mm and a thickness of 0.16 mm. Each of the springs is connected to an anchor frame, which is sandwiched between two anodized aluminum fixtures (four fixtures in total for the two anchors). Among the four aluminum fixtures, two are fixed to the stacked coils and two are free to move along the X direction in Figure 1. The four bolts on the side of the two movable aluminum fixtures can be fastened or loosened to adjust the distance between the movable fixtures and the fixed fixtures. By adjusting the distance, compressive or tensile stress is introduced into the polyimide springs, resulting in the shift of the EVEH's resonance frequency, as shown in Figure 1c. Once the tuning is complete, the position of the movable fixture is fixed by two slidable bolts in the trenches. The other two fixed bolts visible on top of the movable fixture are used to create a firm and strong clamping force for the anchor area of the polyimide springs. During the assembly of the EVEH device, acrylic glue is used between the polyimide anchor and the fixtures to further improve the clamping and facilitate the assembly process. The coils used in the study are a stack of 15 layers of double-sided copper planar coils electroplated on 180 μm-thick polyimide substrates. The flexible coil layers are clamped between two rigid FR4 frames to be electrically interconnected, forming a serially connected high density coils.

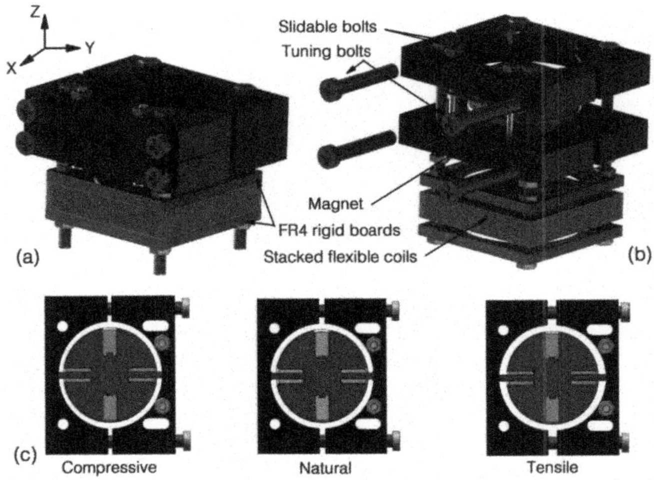

Figure 1. Schematically illustration of (**a**) the assembled EVEH device; (**b**) the EVEH in an exploded view; and (**c**) tuning process of the EVEH.

2.2. Analytical Modeling

For a uniform spring under small deflections, if we assume the axial load is constant, the equation of motion can be obtained by summing the internal shear, moment and inertial forces acting on an element dx in the spring length. Therefore, the equation of motion for the spring under constant axial load F is given by [18]:

$$\frac{\partial^4 V(x,t)}{\partial x^4} + \frac{F}{EI}\frac{\partial^2 V(x,t)}{\partial x^2} + \frac{\rho}{EI}\frac{\partial^2 V(x,t)}{\partial t^2} = 0, \tag{1}$$

where $V(x, t)$ is the deflection of the spring at the position x and time t. F is the axial load. E and I are the Young's modulus and the area moment of inertia of the spring's cross section, respectively and ρ is the mass per length. Assuming that the spring performs harmonic oscillations with an angular frequency ω, the expression for its transversal displacement can be written as:

$$V(x,t) = v(x)\sin(\omega t), \tag{2}$$

Substituting Equation (2) into Equation (1) yields:

$$\frac{\mathrm{d}^4 v(x)}{\mathrm{d}x^4} + k^2\frac{\mathrm{d}^2 v(x)}{\mathrm{d}x^2} - \beta^4 v(x) = 0, \tag{3}$$

where $k^2 = F/EI$ and $\beta^4 = \rho\omega^2/EI$, $v(x)$ describes the oscillation amplitude of the spring at position x. Transforming this equation into a nondimensional form by letting $x = x/l$, $v = v/l$, $k = kl$ and $\beta = \beta l$. The nondimensional equation of motion is then given by:

$$\frac{\mathrm{d}^4 \overline{v}(\overline{x})}{\mathrm{d}\overline{x}^4} + \overline{k}^2\frac{\mathrm{d}^2 \overline{v}(\overline{x})}{\mathrm{d}\overline{x}^2} - \overline{\beta}^4 \overline{v}(\overline{x}) = 0 \tag{4}$$

The solution to this differential equation is given by:

$$\overline{v}(\overline{x}) = A\cos\mathrm{h}(\alpha_1\overline{x}) + B\sin\mathrm{h}(\alpha_1\overline{x}) + C\cos(\alpha_2\overline{x}) + D\sin(\alpha_2\overline{x}), \tag{5}$$

where A, B, C and D are constants that can be obtained from certain boundary conditions and the constants α_1 and α_2 are given by:

$$\begin{aligned} \alpha_1 &= \sqrt{-\frac{\overline{k}^2}{2} + \sqrt{\frac{\overline{k}^4}{4} + \overline{\beta}^4}} \\ \alpha_2 &= \sqrt{\frac{\overline{k}^2}{2} + \sqrt{\frac{\overline{k}^4}{4} + \overline{\beta}^4}} \end{aligned} \tag{6}$$

The spring-magnet structure in this work can be modeled as a clamped-guided beam with tip mass, as shown in Figure 2. The equivalent tip mass is half of the mass of the magnet and reinforcement structure. The boundary conditions of the spring are given as follows:

$$\begin{aligned} &\overline{v}(0) = 0 \\ &\overline{v}'(0) = 0 \end{aligned} \quad \text{Clamped}$$
$$\begin{aligned} &\overline{v}'(1) = 0 \\ &\overline{v}'''(1) + \overline{k}^2\overline{v}'(1) + \tfrac{1}{2}\eta_T\overline{\beta}^4\overline{v}(1) = 0 \end{aligned} \quad \text{Guided with tip mass}, \tag{7}$$

where η_T is the ratio between the tip mass M_T to the spring mass $M = \rho l$.

$$\begin{aligned} &-\eta_T\overline{\beta}^4\alpha_1 + \tfrac{1}{2}\eta_T\overline{\beta}^4\left(\frac{\alpha_2^2-\alpha_1^2}{\alpha_2}\right)\sinh(\alpha_1 l)\sin(\alpha_2 l) \\ &+\eta_T\overline{\beta}^4\alpha_1\cosh(\alpha_1 l)\cos(\alpha_2 l) + \left(\alpha_1^2\alpha_2^2 + \alpha_1^4\right)\sinh(\alpha_1 l)\cos(\alpha_2 l) \\ &+\left(\alpha_1^3\alpha_2 + \alpha_1\alpha_2^3\right)\cosh(\alpha_1 l)\sin(\alpha_2 l) = 0 \end{aligned} \tag{8}$$

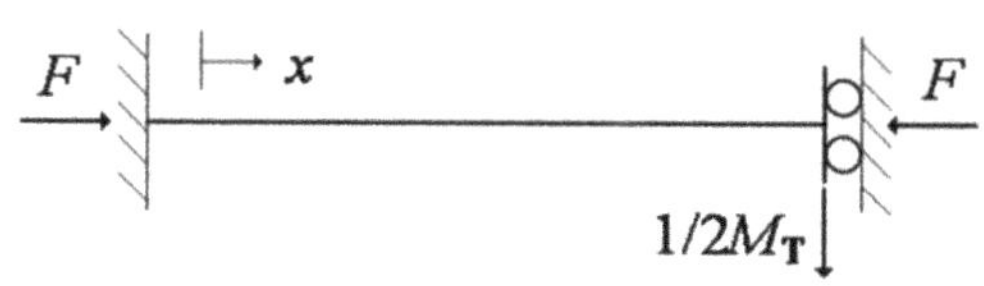

Figure 2. Schematic diagram of the clamped-guided spring with tip mass.

The design and structural parameters of the proposed EVEH is given in Table 1. By using these values, the analytical solution of Equation (8) can be obtained. The normalized frequency ω/ω_0 as a function of normalized axial load F/F_{cr} is shown in Figure 3. When the normalized axial load is 0, the normalized frequency of the device is defined as 1. Under compressive load (F/F_{cr} is positive), ω/ω_0 decreases as the compressive load increases. When the critical buckling point (F/F_{cr} = 1) is reached, ω/ω_0 drops to zero theoretically. Under tensile load (F/F_{cr} is negative), ω/ω_0 increases as the tensile load increases. The maximum value of ω/ω_0 is 1.406, at a F/F_{cr} value of -1. This indicates that when the critical buckling load is applied, the resonance frequency of the EVEH device is 40.6% of the natural resonance frequency of the device. In another word, by adjusting the axial load in the spring, a normalized frequency tuning range of 40.6% with respect to the natural resonance frequency of the EVEH device can be achieved.

Table 1. Design parameters of the EVEH.

Parameter	Description	Value
E	Young's modulus	3.2×10^9 Pa
I	Area moment of inertia, $I = bh^3/12$	3.4×10^{-16} m^4
l	Length of the spring	7.0×10^{-3} m
b	Width of the spring	1.0×10^{-3} m
h	Thickness of the spring	1.6×10^{-4} m
ρ	Weight per length	2.2×10^{-4} kg/m
M_T	Weight of the tip mass	5.5×10^{-3} kg
F_{cr}	Critical buckling load, $F_{cr} = \pi^2 EI/12$	2.2×10^{-1} N

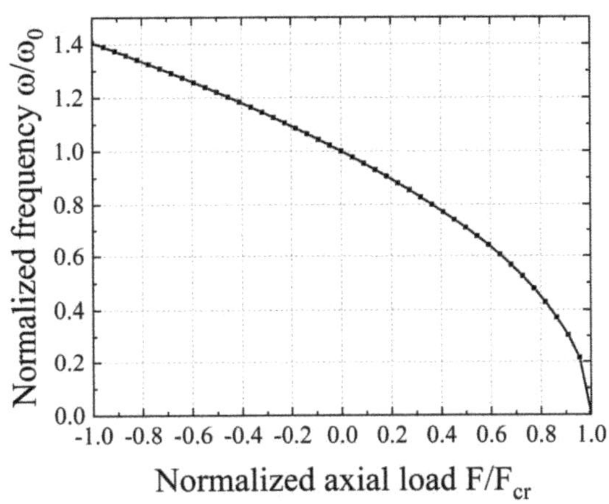

Figure 3. The analytical normalized frequency ω/ω_0 as function of normalized axial load F/F_{cr}.

2.3. Finite Element Method (FEM) Modeling

A FEM modal analysis of the EVEH is performed with Comsol Multiphysics. In the model, the anchoring frame structure is simplified by applying fixed constrains to the end of the springs. This simplification is valid as the springs are fixed at the end when being clamped between two rigid frames. Different axial loads are asserted by specifying different axial displacements of the end of the springs. The material parameters of the polyimide used in the simulation are Young's Modulus of 3.2 GPa, Poisson ratio of 0.34. The four primary vibrational modes of the EVEH are shown in Figure 4: out-of-plane twisting mode around the x-axis at 25.07 Hz, out-of-plane torsional mode around the z-axis at 82.08 Hz, the out-of-plane piston motion mode at 118.78 Hz and in-plane twisting mode around the y-axis at 194.04 Hz. In spite of being the 3rd resonance mode, the piston motion mode will be the dominant resonance mode during operation as the entire EVEH device is subjected to a uniform acceleration vertically (Y-direction in Figure 4). Slight torsional movement may be combined the piston motion movement of the proof mass due to the slight geometrical asymmetry induced by the assembly process of the device, e.g., the off-centered position of the magnet induced by the gluing process.

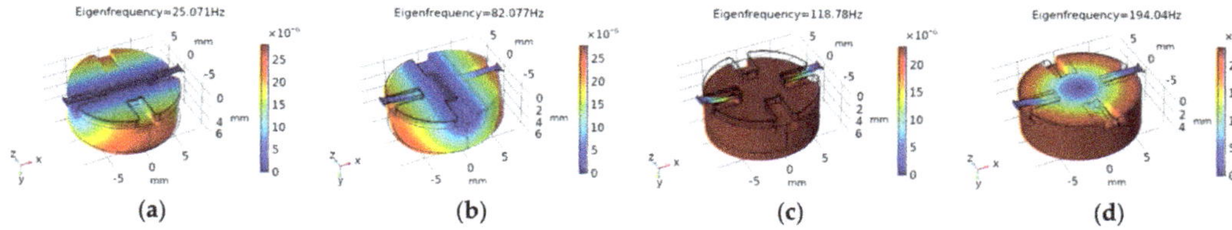

Figure 4. FEM Modal analysis of the EVEH. (**a**) twisting around the x-axis (25.07 Hz); (**b**) twisting around the z-axis (82.08 Hz); (**c**) out-of-plane piston (118.78 Hz); (**d**) in-plane twisting around the y-axis (194.04 Hz).

For the piston mode of the EVEH device, the effect of the axial stress on the resonance frequency is studied. The resonance frequency as a function of the relative fixture distance Δd is shown in Figure 5. The resonance frequency of the device with no stress in the spring is 118.78 Hz. When tensile force ($\Delta d > 0$) is applied, the resonance frequency increases with the increase of Δd. In the simulation, a maximum resonance frequency of 234.46 Hz is obtained with Δd of 0.6 mm. When compressive stress ($\Delta d < 0$) is applied, the resonance frequency decreases rapidly with the increase of stress. It is difficult to predict the influence of spring buckling on the resonance behavior of the EVEH. The simulated resonance frequency of the EVEH drops to 0 at the critical buckling point of $\Delta d = -0.146$ mm. However, it has been shown that buckling stiffens the spring because it changes the resonant movement of spring from dynamic bending to dynamic stretching [19]. Therefore, as Δd further decreases beyond the buckling mode, the spring-stiffening effect originated from the tensile stress induced by buckling will begin to dominate and the resonance frequency will rise accordingly.

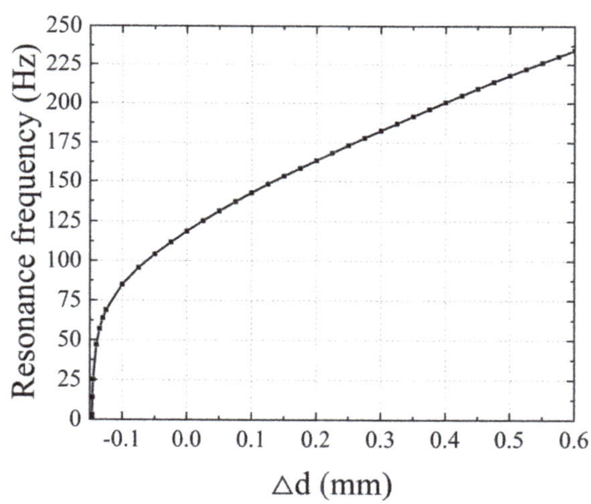

Figure 5. The FEM resonance frequency as function of the relative fixture distance Δd.

3. Experimental

Figure 6 shows the measurement setup for characterizing the EVEH. The setup is capable of generate different vibration conditions while simultaneously measuring the output voltage of the EVEH device. The assembled EVEH is tuned to different resonance frequency and then systematically characterized when being excited by a piezo shaker. The sinusoidal actuation signal of the shaker is generated by a signal generator and amplified by a power amplifier. The proposed EVEH has FR4 rigid PCB board at bottom which facilitates the mounting process. The acceleration applied to the EVEH is measured by an accelerometer installed between the EVEH and the shaker. The signals of the EVEH and the accelerometer are measured simultaneously using a NI USB-6211 data acquisition board. A 4th-order Butterworth low-pass filter is applied to the measured output of the EVEH to filter the high frequency noise. The filter is implemented by using the LabVIEW software and has a cut-off frequency of 1000 Hz which is far beyond the resonance frequency of the EVEH.

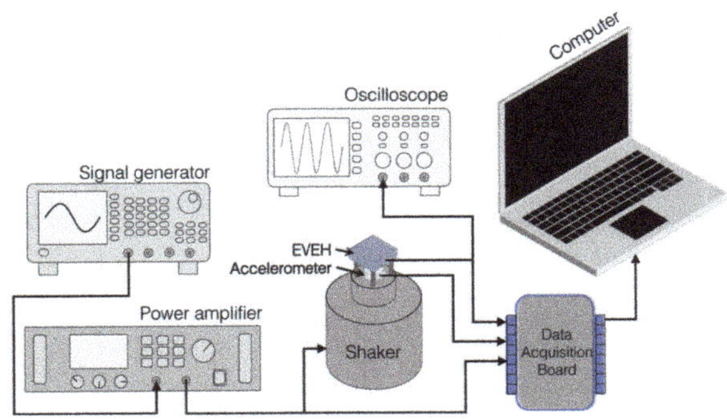

Figure 6. Schematic diagram of the measurement setup.

4. Results and Discussions

This section presents the characterization results of the fabricated tunable EVEH device. As the main goal of the presented work is to demonstrate the frequency tuning capability of the proposed EVEH device, only open-loop performance of the EVEH device in the time and frequency domain is presented, while the impedance matching and closed-loop performance of the EVEH device is not reported in this work.

4.1. Manufactured EVEH Device

The photo of the assembled EVEH device is shown in Figure 7a. At the bottom of the device is the stacked flexible coils sandwiched between two rigid FR4 frames. Fifteen layers of 180 µm-thick flexible coils are included in the device shown in the photo. The disc magnet is mounted above the coil stack and suspended by two straight polyimide springs. The anchor of one spring is clamped between two fixed aluminum fixtures, while the anchor of the other spring is clamped between two movable fixtures. The surfaces of the fixtures are black anodized for a better surface protection. The fixtures in the photo are in a slightly compressed state and, thus, the springs are buckled downwards. The magnet in this work is a circular N50 NdFeB magnet with a diameter of 15 mm and a thickness of 4 mm. Figure 7b shows the magnet, the springs and anchors of the EVEH. The magnet is glued onto a reinforcement structure attached to the top surface of the polyimide layer, in order to create space for the springs to oscillate below the magnet and minimize the footprint of the device. Two independent anchors are connected to the end of the springs. When the springs are in a natural state (neither stretched nor compressed), the gap between the magnet and the coil is 4.8 mm. This distance will decrease when the spring is buckled downwards by the compressive stress introduced during the tuning process. Figure 7c shows the photo of a layer of the flexible coils. The coil layer consists of spiral copper wires electroplated on the two sides of a polyimide layer. The patterns of the coils are identical on the two sides but 180 flipped degrees. The winding directions of the coils on the two sides are opposite and are interconnected by the via in the center of the layer. The electrodes at the edge are used to connect coils on different layers. By clamping many coil layers together with 180-degree rotation, a high-density coil consisting of many sub-coils connected in series is formed.

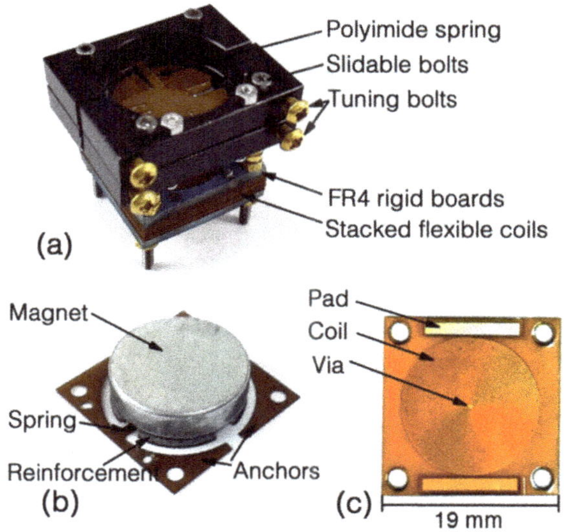

Figure 7. Photos of (**a**) the assembled EVEH device; (**b**) the magnet on the reinforcement structure fixed to the spring layer; and (**c**) a layer of the planar coils.

4.2. Time Domain Response

The lengths of the springs are tuned by adjusting the distance between the two pairs of fixtures, which is realized by tightening or loosening the side tuning bolts. At each distance, the open-circuit output voltage of the EVEH is measured. The applied excitation signal is a sinusoidal vibration with a peak-to-peak value of 2 g. Figure 8 shows the output voltage of the EVEH as a function of time, measured at the resonance frequency of five different relative distances between fixtures (Δd, with respect to distance between the fixture when the spring is in the natural and relaxed state). These 5 distances correspond to the natural state of the spring (Δd = 0 mm), the extreme compressive state of the spring (Δd = −0.3 mm), the extreme tensile state of the spring (Δd = 0.5 mm) and two intermediate stress levels of the springs (Δd = −0.1 mm and Δd = 0.3 mm), respectively. The measured output curves have mostly well-defined sinusoidal characteristics with no distortion or nonlinearity, except the curve at Δd = 0.3 mm. The curve at Δd = −0.3 mm exhibits slight non-symmetric characteristics, as a result of the buckling-induced nonlinearity from elevated compression stress level. The period of the EVEH's output voltage corresponds with the excitation vibration frequency. The peak-to-peak output voltage of the EVEH under natural state (Δd = 0 mm) is 749.8 mV. It is lower than the results of 1.57 V we have reported in another work due to the large coil-magnet distance in this work [17]. As the main goal of this EVEH is to tune the resonance frequency, only open-loop performance of the EVEH device is investigated in this work, while the impedance matching and closed-loop performance of the EVEH device is not detailed. The power produced by this device under natural state (Δd = 0 mm) with a matching resistance of 923 Ω is approximately 19.04 μW.

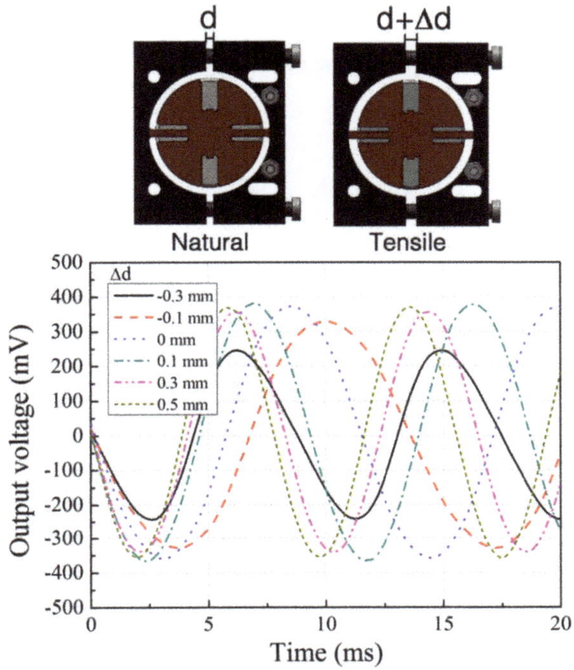

Figure 8. Output voltage of the EVEH versus time, measured at different relative fixture distances Δd.

4.3. Frequency Tuning of the Device

Figure 9 shows the output voltage of the EVEH as a function of the frequency, measured at different fixture distance. All the voltage-frequency curves show typical linear characteristic with symmetric resonance peaks. The total tuning frequency range is 56 Hz (74–130 Hz), which is 64% of the natural resonance frequency of the EVEH (88 Hz). The tuning range of 64% is 3 times higher than the tuning range of 22% realized by an active frequency tuning technique (which consumes energy) of an EVEH with similar resonance frequency (64 Hz) [20]. When tensile stress is induced in the spring (Δd = 0.1–0.5 mm), the resonance peak of the EVEH shifts towards higher frequency, because tensile stress increases the effective spring stiffness. In addition, the influence of the tensile stress on the output voltage amplitude of the EVEH is not obvious while tuning the resonance frequency. When compressive stress is induced in the spring, the resonance frequency first shifts towards lower frequency (Δd = −0.1 mm) and then shifts towards higher frequency (Δd = −0.2 mm and −0.3 mm). This phenomenon will be discussed with respect to Figure 10. The output amplitude first decreases (Δd = −0.1 mm and −0.2 mm) and then increases (Δd = −0.3 mm). Such a characteristic is induced by the combined effect of the increased spring stiffness from spring buckling and decreased coil-magnet distance.

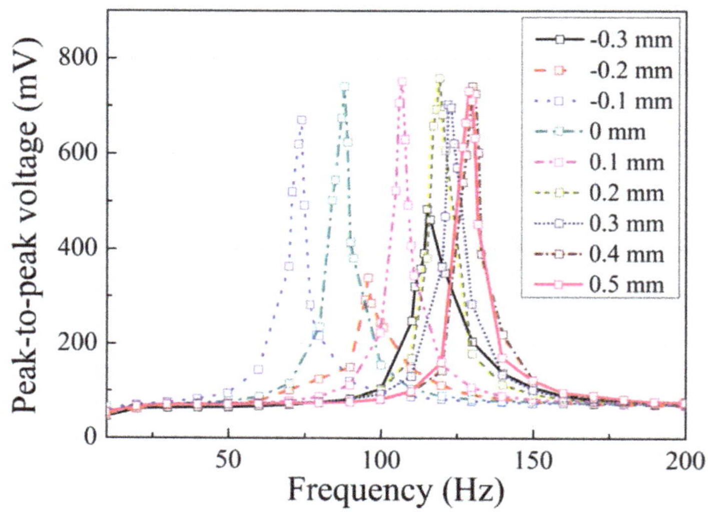

Figure 9. Output peak-to-peak voltage as a function of frequency, measured at different relative fixture distance Δd.

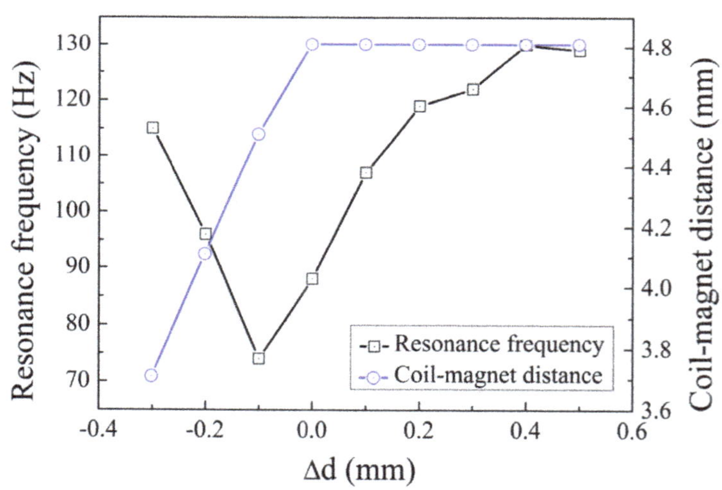

Figure 10. Resonance frequency and coil-magnet distance as a function of relative fixture distance Δd.

Figure 10 shows the resonance frequency and coil-magnet distance as a function of the relative distance Δd between the fixtures. The coil-magnet distance is constant (4.8 mm) when the tensile stress is introduced into the spring (Δd = 0.1–0.5 mm), because no buckling occurs with the presence of tensile stress. The resonance frequency increases from 88 Hz to 130 Hz when Δd increases from 0.1 mm to 0.5 mm. The increase in the resonance frequency is a result of the enhanced effective spring stiffness for the bending mode of the spring. When the compressive stress is introduced into the spring, the coil-magnet distance decreases from 4.8 mm to 3.7 mm. Meanwhile, the resonance frequency first decreases from 88 Hz to 74 Hz and then increases to 115 Hz. The decrease in the resonance frequency is due to the reduced effective spring stiffness by the compressive stress and the subsequent increase in the resonance frequency is induced by the buckling. When the compressive stress increases beyond the buckling point, the excess length of the spring is relaxed by a central deflection, the stress will not increase anymore. Although compressive stress is known to decrease the spring stiffness, the buckling effect induced by the compressive stress is known to increase the effective spring stiffness, due to the dynamic axial force and the dynamic stretching of the springs. This behavior of successive decrease and increase in the resonance frequency induced by compressive stress is identical as the pattern predicted by Bouswtra and Geijselaers [21].

Table 2 lists the comparison between the present work and recent works of tunable energy harvesters based on both band broadening and frequency tuning techniques. For band broadening techniques, the multi-mode devices have rather narrow band width (often less than 10% of their resonance frequency) and the frequency band is the superposition of several narrow frequency bands [8,9]. The frequency tuning range of 64% in this work is larger than the bandwidth of the multi-mode devices. The multi-frequency device often utilizes array structures with different resonance frequency to expand frequency band. Despite the very large frequency band of ~1000 Hz [13]. The very high resonance frequency (~4000 Hz) deteriorates the applicability of such device in most of the ambient vibrations (1–200 Hz). For low frequency vibration energy harvesting, the non-linear [14], bistable [22], magnetostatic [23] wideband energy harvesters have the bandwidth between 5–19%, which is also lower than the relative tuning range of 64% in this work. These three types of devices have the potential to realize very large bandwidth, but each of them must overcome their inherent behavioral limitations before put into real applications, e.g., hysteresis for nonlinear devices, uncertain dynamic behavior under random and low vibrations for bistable devices and limitation of inter-magnet distance. For frequency tuning techniques, one of the most common techniques is to use magnetic force/coupling to tune the effective spring stiffness and, thus, the resonance frequency [20,24,25]. It is a simple and effective method, but its tuning range is limited by the distance between the magnets. Furthermore, the magnetic force increases rapidly when the distance between the magnets is small, giving rise to very limited output level at high frequency regime. The frequency tuning process can also be realized by adjusting the stress in the beam, which is often used in the piezoelectric energy harvesters [26–28]. Despite the effectiveness of tuning performance, the tuning structure used in these devices are often bulky, e.g., 1700 cm^3 for a 6–62 Hz frequency range. Few works have been reported on the stress modulation in the spring of electromagnetic energy harvesters. In [29], the reported VEH device achieves a 65% frequency tuning range by applying an axial load on an aluminum spring with a central point mass. In this work, similar relative frequency tuning range is realized with a compact device with 28 times smaller volume compared to the VEH in [29]. It is worth mentioning that besides the common methods listed in Table 2, there are also other novel approaches to broaden the bandwidth of energy harvesters, such as using rotational pendulum [30], anti-resonance structure [31] and implementing the VEH in a non-resonating principle [32]. Compared with most of the works listed in Table 2, the tunable EVEH in this work has higher absolute and relative frequency tuning range, with a relative compact volume. However, before the proposed tunable EVEH can be used in real-world applications, the nonlinearity in the tuning performance must be addressed.

Table 2. Comparison between tunable energy harvesters.

	Reference	Technology	Size/cm^3	Absolute Frequency Range/Hz	Relative Frequency Range/%
Band Broadening	[8]	Multi-mode device	1	71–74 105–107 111–114	4 2 3
	[9]	Multi-mode device	3.18	345–395 900–980 1130–1230	14 9 8
	[13]	Multi-frequency device, 40 cantilevers	1.40	3500–4500	25
	[13]	Multi-frequency device, 35 cantilevers	1.40	4200–5000	17
	[14]	Nonlinear device	9	94.1–98.9	5
	[22]	Bistable device	2.97	30–35	15
	[23]	Magnetostatic coupling device	68.9	45.5–55.5	19
Frequency Tuning	[24]	Combination of linear springs and magnetic nonlinearity	61	2–7	110
	[20]	Magnetic coupling adjustment of the spring stiffness	1.12	64–78	22
	[25]	Magnetic coupling adjustment of the spring stiffness	1.8	409–516	25
	[26]	Stress modulation of piezoelectric springs	1700	6–62	191
	[27]	Stress modulation of piezoelectric cantilever	54	292–380 440–460	22 4
	[28]	Stress modulation of a piezoelectric spring	60	200–250	24
	[29]	Stress modulation of an aluminum spring	280	16–46	65
	This work	Stress modulation of polymeric springs	9.95	74–130	64

5. Conclusions

In this paper, we report an electromagnetic vibrational energy harvester with tunable resonance frequency. The resonance frequency of the EVEH is tuned by adjusting the axial stress introduced into the springs, which is realized by physically pulling and pushing the springs via a pair of metallic fixtures. The total tuning frequency range is 56 Hz (74–130 Hz), which is 64% of the natural resonance frequency of the EVEH (88 Hz). The tensile stress increases the resonance frequency of the EVEH, while the compressive stress firstly reduces the resonance frequency and then increases the resonance frequency due to buckling. In the future, new spring design might be implemented to improve the nonlinear relationship between the tuned resonance frequency and the tuning parameter (Δd). In addition, the tuning range of the resonance frequency might be further increased by optimizing the spring and fixture structure.

Author Contributions: Conceptualization, Y.L. and K.T.; Data curation, X.W.; Formal analysis, D.Q.; Investigation, C.Z., Q.C. and K.T.; Methodology, C.Z. and Q.C.; Writing—review & editing, Y.L. All authors have read and agreed to the published version of the manuscript.

Funding: This work was funded by the National Key Research and Development Program of China (2018YFF01010902) and the National Natural Science Foundation of China (51705405).

Institutional Review Board Statement: Not applicable.

Informed Consent Statement: Not applicable.

Conflicts of Interest: The authors declare no conflict of interest.

References

1. Priya, S.; Inman, D.J. *Energy Harvesting Technologies;* Springer Science+Business Media, LLC: New York, NY, USA, 2008; p. 524.
2. Tao, K.; Wu, J.; Tang, L.; Xia, X.; Lye, S.W.; Miao, J.; Hu, X. A novel two-degree-of-freedom MEMS electromagnetic vibration energy harvester. *J. Micromech. Microeng.* **2016**, *26*, 035020. [CrossRef]
3. Tang, L.; Yang, Y.; Soh, C. Toward Broadband Vibration-based Energy Harvesting. *J. Intell. Mater. Syst. Struct.* **2010**, *21*, 1867–1897. [CrossRef]
4. Roundy, S.; Leland, E.S.; Baker, J.; Carleton, E.; Reilly, E.; Lai, E.; Otis, B.; Rabaey, J.M.; Wright, P.K.; Sundararajan, V. Improving power output for vibration-based energy scavengers. *IEEE Pervasive Comput.* **2005**, *4*, 28–36. [CrossRef]
5. Mitcheson, P.D.; Toh, T.T.; Wong, K.H.; Burrow, S.G.; Holmes, A.S. Tuning the Resonant Frequency and Damping of an Electromagnetic Energy Harvester Using Power Electronics. *IEEE Trans. Circuits Syst. II: Express Briefs* **2011**, *58*, 792–796. [CrossRef]
6. Adams, S.G.; Bertsch, F.M.; Shaw, K.A.; Hartwell, P.G.; Moon, F.C.; MacDonald, N.C. Capacitance based tunable resonators. *J. Micromech. Microeng.* **1998**, *8*, 15. [CrossRef]
7. Roundy, S.; Zhang, Y. Toward self-tuning adaptive vibration-based microgenerators. *Smart Mater. Nano- Micro-Smart Syst.* **2005**, *5649*, 12.
8. Ching, N.N.H.; Wong, H.Y.; Li, W.J.; Leong, P.H.W.; Wen, Z. A laser-micromachined multi-modal resonating power transducer for wireless sensing systems. *Sens. Actuator A-Phys.* **2002**, *97–98*, 685–690. [CrossRef]
9. Yang, B. Electromagnetic energy harvesting from vibrations of multiple frequencies. *J. Micromech. Microeng.* **2009**, *19*, 035001. [CrossRef]
10. Chen, J.; Chen, D.; Yuan, T.; Chen, X. A multi-frequency sandwich type electromagnetic vibration energy harvester. *Appl. Phys. Lett.* **2012**, *100*, 213509. [CrossRef]
11. Liu, H.; Soon, B.W.; Wang, N.; Tay, C.J.; Quan, C.; Lee, C. Feasibility study of a 3D vibration-driven electromagnetic MEMS energy harvester with multiple vibration modes. *J. Micromech. Microeng.* **2012**, *22*, 125020. [CrossRef]
12. El-Hebeary, M.M.R.; Arafa, M.H.; Megahed, S.M. Modeling and experimental verification of multi-modal vibration energy harvesting from plate structures. *Sens. Actuator A-Phys.* **2013**, *193*, 35–47. [CrossRef]
13. Sari, I.; Balkan, T.; Kulah, H. An electromagnetic micro power generator for wideband environmental vibrations. *Sens. Actuator A-Phys.* **2008**, *145–146*, 405–413. [CrossRef]
14. Soliman, M.S.M.; Abdel-Rahman, E.M.; El-Saadany, E.F.; Mansour, R.R. A wideband vibration-based energy harvester. *J. Micromech. Microeng.* **2008**, *18*, 115021. [CrossRef]
15. Nguyen, D.S.; Halvorsen, E.; Jensen, G.U.; Vogl, A. Fabrication and characterization of a wideband MEMS energy harvester utilizing nonlinear springs. *J. Micromech. Microeng.* **2010**, *20*, 125009. [CrossRef]
16. Dhiman, M.; Andreas, A.; Saibal, R. A nonlinear stretching based electromagnetic energy harvester on FR4 for wideband operation. *Smart Mater. Struct.* **2015**, *24*, 015013.
17. Li, Y.; Cao, Q.; Zhang, W.; Zhang, Y.; Cao, J. A miniaturized electromagnetic energy harvester with volt-level output based on stacked flexible coils. *Smart Mater. Struct.* **2018**, *27*, 115040.
18. Shaker, F.J. *Effect of axial load on mode shapes and frequencies of beams;* NASA Tech. Note: Washington, DC, USA, 1975.
19. Holst, G.L.; Teichert, G.H.; Jensen, B.D. Modeling and Experiments of Buckling Modes and Deflection of Fixed-Guided Beams in Compliant Mechanisms. *J. Mech. Des.* **2011**, *133*, 051002. [CrossRef]
20. Garcia, I.A.; Zhu, D.; Tudor, J.; Beeby, S. Autonomous Tunable Energy Harvester. In Proceedings of the Power MEMS 2009, Washington, DC, USA, 1–4 December 2009.
21. Bouwstra, S.; Geijselaers, B. On the resonance frequencies of microbridges. In Proceedings of the TRANSDUCERS'91: 1991 International Conference on Solid-State Sensors and Actuators. Digest of Technical Papers, San Francisco, CA, USA, 24–27 June 1991.
22. Podder, P.; Amann, A.; Roy, S. A bistable electromagnetic micro-power generator using FR4-based folded arm cantilever. *Sens. Actuator A-Phys.* **2015**, *227*, 39–47.
23. Xing, X.; Lou, J.; Yang, G.M.; Obi, O.; Driscoll, C.; Sun, N.X. Wideband vibration energy harvester with high permeability magnetic material. *Appl. Phys. Lett.* **2009**, *95*, 134103. [CrossRef]
24. Chen, Y.; Pollock, T.E.; Salehian, A. Analysis of compliance effects on power generation of a nonlinear electromagnetic energy harvesting unit; theory and experiment. *Smart Mater. Struct.* **2013**, *22*, 94027. [CrossRef]
25. Podder, P.; Constantinou, P.; Amann, A.; Roy, S. Frequency adjustable MEMS vibration energy harvester. *J. Phys. Conf. Ser.* **2016**, *757*, 12037. [CrossRef]
26. Dehghan Niri, E.; Salamone, S. A passively tunable mechanism for a dual bimorph energy harvester with variable tip stiffness and axial load. *Smart Mater. Struct.* **2012**, *21*, 125025. [CrossRef]

27. Eichhorn, C.; Goldschmidtboeing, F.; Woias, P. Bidirectional frequency tuning of a piezoelectric energy converter based on a cantilever beam. *J. Micromech. Microeng.* **2009**, *19*, 094006. [CrossRef]
28. Leland, E.S.; Wright, P.K. Resonance tuning of piezoelectric vibration energy scavenging generators using compressive axial preload. *Smart Mater. Struct.* **2006**, *15*, 1413–1420. [CrossRef]
29. Yildirim, T.; Ghayesh, M.H.; Li, W.; Alici, G. A Nonlinearly Broadband Tuneable Energy Harvester. *J. Dyn. Syst. Meas. Control-Trans. ASME* **2017**, *139*, 011008. [CrossRef]
30. Ambrożkiewicz, B.; Litak, G.; Wolszczak, P. Modelling of Electromagnetic Energy Harvester with Rotational Pendulum Using Mechanical Vibrations to Scavenge Electrical Energy. *Appl. Sci.* **2020**, *10*, 671. [CrossRef]
31. Foong, F.M.; Thein, C.K.; Ooi, B.L.; Yurchenko, D. Increased power output of an electromagnetic vibration energy harvester through anti-phase resonance. *Mech. Syst. Signal Process.* **2019**, *116*, 129–145. [CrossRef]
32. Tao, K.; Chen, Z.; Yi, H.; Zhang, R.; Shen, Q.; Wu, J.; Tang, L.; Fan, K.; Fu, Y.; Miao, J.; et al. Hierarchical Honeycomb-Structured Electret/Triboelectric Nanogenerator for Biomechanical and Morphing Wing Energy Harvesting. *Nano-Micro Lett.* **2021**, *13*, 1–16. [CrossRef]

micromachines

Article

All-in-One High-Power-Density Vibrational Energy Harvester with Impact-Induced Frequency Broadening Mechanisms

Yongqi Cao [1,2,†], Weihe Shen [3,†], Fangzhi Li [1], Huan Qi [4], Jiaxiang Wang [1], Jianren Mao [3], Yang Yang [2] and Kai Tao [1,2,*]

1 Ministry of Education Key Laboratory of Micro and Nano Systems for Aerospace, Northwestern Polytechnical University, Xi'an 710072, China; caoyq@mail.nwpu.edu.cn (Y.C.); 2017300086@mail.nwpu.edu.cn (F.L.); jiaxiang@mail.nwpu.edu.cn (J.W.)
2 Research & Development Institute in Shenzhen, Northwestern Polytechnical University, Shenzhen 518057, China; yang_yang@cqu.edu.cn
3 Science and Technology on Space Physics Laboratory, China Academy of Launch Vehicle Technology, Beijing 100076, China; weihenihao@gmail.com (W.S.); maoren_99@163.com (J.M.)
4 Beijing Institute of Astronautical Systems Engineering, China Academy of Launch Vehicle Technology, Beijing 100076, China; qi_huan@sina.com
* Correspondence: taokai@nwpu.edu.cn
† These authors contributed equally to this work.

Abstract: This paper proposes an electrostatic-piezoelectric-electromagnetic hybrid vibrational power generator with different frequency broadening schemes. Both the nonlinear frequency broadening mechanisms and the synergized effect of the electrostatic-piezoelectric-electromagnetic hybrid structures are investigated. The structure and performance of the composite generator are optimized to improve the response bandwidth and performance. We propose that the electrostatic power generation module and the electromagnetic power generation module be introduced into the cantilever beam to make the multifunctional cantilever beam, realizing small integrated output loss, high output voltage, and high current characteristics. When the external load of the electrostatic power generation module is 10 kΩ, its peak power can reach 3.6 mW; when the external load of the piezoelectric power generation module is 2 kΩ, its peak power is 2.2 mW; and when the external load of the electromagnetic power generation module is 170 Ω, its peak power is 0.735 mW. This means that under the same space utilization, the performance is improved by 90%. Moreover, an energy management circuit (ECM) at the rear end of the device is added, through the energy conditioning circuit, the device can directly export a 3.3 V DC voltage to supply power to most of the sensing equipment. In this paper, the hybrid generator's structure and performance are optimized, and the response bandwidth and performance are improved. In general, the primary advantages of the device in this paper are its larger bandwidth and enhanced performance.

Keywords: vibration energy harvesting; all-in-one; frequency broadening

Citation: Cao, Y.; Shen, W.; Li, F.; Qi, H.; Wang, J.; Mao, J.; Yang, Y.; Tao, K. All-in-One High-Power-Density Vibrational Energy Harvester with Impact-Induced Frequency Broadening Mechanisms. *Micromachines* **2021**, *12*, 1083. https://doi.org/10.3390/mi12091083

Academic Editor: Kenji Uchino

Received: 28 July 2021
Accepted: 5 September 2021
Published: 8 September 2021

1. Introduction

Over the past few decades, tremendous advances have been made in microelectronic systems, with devices becoming smaller and requiring less energy [1]. However, limited by the service life and energy density of the traditional batteries, the power supply scheme of these systems is still a challenge. Therefore, researchers hope to design some devices to harvest the energy in the environment to power these microelectronic systems autonomously. Meanwhile, these low-power electronic devices pose a challenge to cheap, flexible, portable, and sustainable energy resources [2–9]. Therefore, many researchers are devoted to the demand of these microelectronic systems, among which vibration energy harvesting technology has become a research hotspot [10–21].

The energy conversion mechanism based on ambient vibration is electrostatic (ES), electromagnetic (EM), piezoelectric (PE), and triboelectric [22–27]. In terms of the electrostatic mechanism, Naruse et al. [28] present a low-frequency electrostatic micro-vibrational energy harvester (VEH) supported by microspheres. The result shows that the output power can reach 40 μW under the excitation of 2 Hz and 0.4 g vibration. Nobuhide Kasagi et al. [29] present an microelectromechanical systems (MEMS) electret generator with nonlinear spring. Additionally, the power output of 1 μW has been obtained at 63 Hz. Ugur Erturun et al. [30] present a vibration energy harvester using push–pull electrostatic conversion. The stored energy of ~900 μJ and the power of ~15 μW are obtained by charging a capacitor for around a minute. Additionally, the output voltage can reach 318 V. James E. West et al. [31] demonstrate an energy harvester that combines a piezoelectric nanogenerator and an electret-based electrostatic generator. The maximum peak–peak voltage output of the electrostatic part is 500 V at 20 Hz. The short-circuit current output of the electrostatic part is 1.1 μA at 10 Hz. In these studies, the electrostatic energy harvesting method can output a large voltage, but the current generated is small.

In terms of the electromagnetic mechanism, Kulkarni et al. [32] present a miniature electromagnetic VEH with the help of MEMS technology. By changing the structure, the maximum output voltage and power of the device can reach 950 mV and 85 μW. Külah et al. [33] present an electromagnetic VEH that converts a low-frequency vibration into a higher-frequency vibration through frequency conversion technology. The maximum output voltage and power obtained by VEH at a natural frequency of 64 Hz are 6 mV and 120 nW, respectively. Peihong Wang [34] uses MEMS technology to make a new electromagnetic VEH. When the acceleration is 0.8 g and the frequency is 280.9 Hz, the maximum output voltage and power are 101 mV and 19.5 μW, respectively. Kankana Paul et al. [35] present the design and performance of fully integrated EM vibration energy harvesters on the scale of microelectromechanical systems (MEMS). The device produces a power density as high as 52 $\mu W/cm^3$ at 1 g acceleration. The nonlinear counterpart enhances the bandwidth almost six times to 25 Hz at the cost of reduced power density. In these studies, the voltage performance of the small-size electromagnetic energy generator in the low-frequency environment is not very prominent, but the current generated using the electromagnetic energy harvesting method is large.

Additionally, in terms of the piezoelectric mechanism, a small piezoelectric cantilever beam type linear narrow-band generator is proposed by Bai et al. [36], which can acquire the average power of 50 and 20 μW, respectively, when placed on the arm and top of the human body, and the power density is 0.35 and 0.14 $\mu W/mm^3$, respectively. However, linear narrow-band generators cannot adapt to variable environmental vibrations at any time and cannot provide a stable output. Therefore, Shahruz et al. [37] proposed an array piezoelectric broadband generator. The different natural frequencies of the array cantilever beams increase the bandwidth of the generator. Xue et al. [38] also proposed a piezoelectric generator with a bimorph cantilever beam array, which uses different resonance frequencies caused by different wafer thicknesses to increase the bandwidth of the generator. However, the average power of the array generator is low. Leland et al. [39] proposed a clamped beam adjustable resonant generator with axial compression preload. This generator can change the beam stiffness to change the natural frequency and increase the bandwidth. However, the adjustable resonant generator cannot respond quickly to changes in excitation. In addition to the above two types of generators that achieve broadband, Li et al. [40] proposed a hardened extrusion-type nonlinear generator that can harvest energy at a lower vibration level. Alireza Rezaniakolaie et al. [41] proposed a trapezoid harvesting multi-beam structure with macro-fiber-composite toward the energy conversion enhancement of piezoelectric energy harvesters from wideband excitation signals. The power generation by this harvester is 84% greater than a common multi-beam design at a 47%-reduced volume, resulting in a 160% power density improvement. Additionally, Weiqun Liu et al. [42] proposed a nonlinear generator with curved surface fixtures, effectively increasing the bandwidth and power.

In general, there are many methods to implement broadband generators, but the nonlinear generator is a more effective solution due to its large bandwidth when carrying out energy harvesting. Therefore, this paper proposes a hybrid nonlinear generator with vibration energy harvesting technologies. On the one hand, we adopt the curved fixture structure to increase the bandwidth. On the other hand, the electrostatic [43–48] generator module and the electromagnetic module are integrated with the piezoelectric generator module to improve the output power of the nonlinear generator. We propose that the electrostatic power generation module and the electromagnetic power generation module be introduced into the cantilever beam to make the multifunctional cantilever beam, realizing small integrated output loss, high output voltage, and high current characteristics.

2. The Conception of Device

This paper designs a hybrid nonlinear generator that combines three power generation methods. The structure design of the power generation is a multifunctional cantilever beam structure composed of the piezoelectric power generation module, a partial structure of the electrostatic power generation module, and a partial structure of the electromagnetic power generation module. As shown in Figure 1, this paper integrates the piezoelectric power generation module, the electrostatic power generation module, the electromagnetic power generation module, and the energy conditioning circuit module based on the LTC3588 chip into a package.

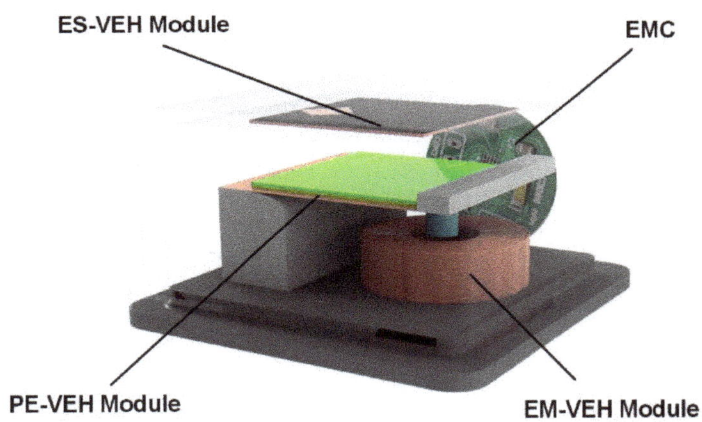

Figure 1. Overall diagram of the electrostatic-piezoelectric-electromagnetic hybrid vibrational power generator: ES-VEH module is the electrostatic power generation module, PE-VEH module is the piezoelectric power generation module, EM-VEH module is the electromagnetic power generation module, and EMC is the energy management circuit module.

The power generation principle of the three power generation modules is shown in Figure 2. The piezoelectric power generation module in this paper is mainly based on the positive piezoelectric effect of piezoelectric materials. Its working principle is as shown in Figure 2a, if the pressure to the direction of the polarization is applied to the piezoelectric chip, due to the compression and deformation of the piezoelectric chip, the distance between the positive and negative charges in the piezoelectric layer will decrease and the polarization strength of the chip will be weakened, which causes a part of the free electrons on the surface electrode of the piezoelectric layer to be released and a discharge phenomenon occurs. After the pressure is removed, the distance between the positive and negative charges in the chip will increase, and the polarization strength will also increase. At this time, a part of the free charges will be adsorbed on the surface of the upper and lower electrodes of the piezoelectric layer and a charging phenomenon will occur.

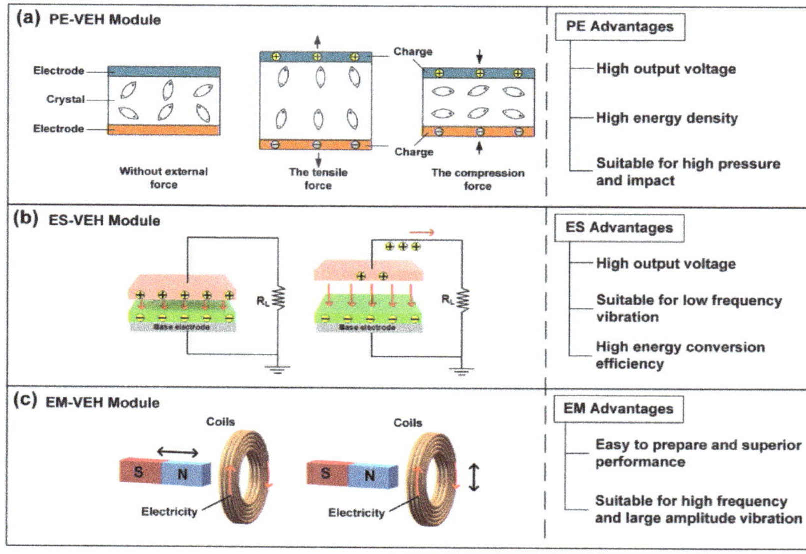

Figure 2. The principle and advantages of the three power generation modules.

In the actual working process, the device's piezoelectric layer constantly repeats the charging and discharging process under the influence of the alternating forces provided by the external vibration environment in order to realize the conversion of mechanical energy into electrical energy. The working principle of the electrostatic power generation module is to convert external mechanical disturbances into capacitance changes of variable capacitors under a constant bias voltage, which causes the charge flow between the two electrodes, thereby converting mechanical energy into electrical energy under external excitation. The FEP is a dielectric with permanent charge or dipole polarization. It can form a permanent electric field around it; therefore, it can be used as a constant voltage source to provide a bias voltage for the electrostatic power generation module.

The working principle of the electrostatic power generation module based on out-of-plane motion is shown in Figure 2b. Due to the electrostatic induction, the negative charge in the electret material on the movable electrode will generate the corresponding positive charge on the fixed electrode, and the positive charge will flow back and forth with external disturbances, thereby converting the mechanical energy into electric energy.

The working principle of the electromagnetic power generation module in this paper is based on the electromagnetic induction, which converts the change of the magnetic flux in the coil into the induced current. It is generally composed of a permanent magnet, a coil, and an elastic unit. When the outside vibrates, the elastic unit will induce the vibration of the outside, causing relative movement between the magnet and the coil, and the coil will sense the changing magnetic flux to generate the output of the electrical energy. The relative movement between the permanent magnet and the coil can be either the permanent magnet moving without the coil moving, or the coil moving without the permanent magnet moving (Figure 2c). The electromagnetic power generation module of this paper uses the electromagnetic power generation method that the permanent magnet moves, and the coil does not move.

Among them, the piezoelectric power generation module uses an MEMS piezoelectric vibration energy harvester design scheme. The dual-crystal cantilever beam of this module uses a metal substrate piezoelectric film multilayer composite structure preparation process, which mainly includes the surface polishing of the material to be bonded, the hot-press bonding of conductive silver glue, the mechanical thinning of the piezoelectric layer, metal electrode sputtering, laser cutting, fixing of the cantilever beam, and the lead process. To enhance the electrical performance, the piezoelectric power generation module adopts the bonding and thinning technology based on the intermediate layer to fabricate the film with excellent piezoelectric properties. At the same time, the mass block is attached to the free end of the cantilever beam, thus, further improving the output performance.

In the piezoelectric power generation module, the PZT piezoelectric material is used as the piezoelectric functional layer. The piezoelectric cantilever beam is used as the main structure of the energy harvester, and a mass block is added at the free end of the beam. The electrostatic power generation module is composed of a movable electrode and a fixed electrode. The fixed electrode is a Cu material layer attached to the top of the internal side of the package shell. The movable electrode is the pre-charged electret material film (FEP) attached to the Cu material layer and bonded to the piezoelectric ceramic generating layer. The electromagnetic power generation module is composed of an inductance coil and a cylindrical permanent magnet. The inductance coil is placed in the center of the package shell base, and the permanent magnet is glued under the cantilever beam to provide a magnetic field for the electromagnetic power generation module.

In this way, the piezoelectric power generation module, the electret layer of the electrostatic power generation module, and the permanent magnet of the electromagnetic power generation module are integrated into a multifunctional cantilever beam that incorporates three power generation methods. The manufacturing process of the multifunctional cantilever beam is shown in Figure 3.

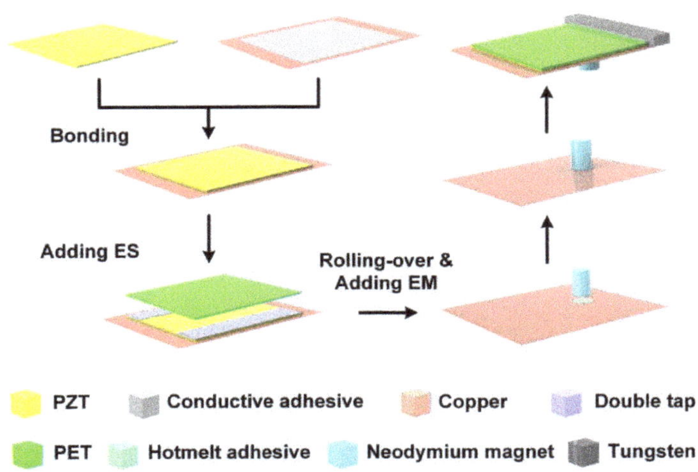

Figure 3. The manufacturing process of the multifunctional cantilever beam.

The energy conditioning circuit mainly consists of LTC3588-1, which integrates a full-wave bridge rectifier with low loss and a buck converter with high efficiency. The package is a 3D printed shell made of resin material. The entire device collects vibration energy in the environment by piezoelectric and electrostatic power generations. It then converts into a well-regulated output to power application microcontrollers, sensors, data converters, and wireless transmission components.

Figure 4 shows the movement of the three power generation modules when the device is subjected to vibration. When affected by the external vibration, the free end mass of the multifunctional cantilever beam vibrates the cantilever beam up and down. The piezoelectric crystal produces the piezoelectric effect. The distance between the FEP movable electrode layer on the piezoelectric layer and the fixed electrode on the top of the shell changes, and thus the electrostatic effect is generated. The permanent magnet at the bottom of the cantilever beam starts to move up and down in the inner ring of the inductor coil to generate electromagnetic induction to complete the power generation process. At the same time, this article explores the nonlinear effect, and uses the bending fixture to broaden the frequency band. The advantages of the bending fixture will be given in the experimental section.

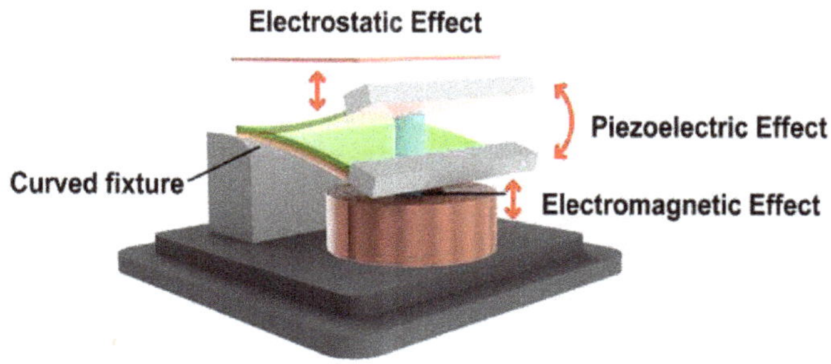

Figure 4. The movement of the three power generation modules.

The vibration of the multifunctional cantilever beam causes the piezoelectric, electrostatic, and electromagnetic power generation modules to start working at the same time. The power generation of the three power generation modules is based on the movement of the multifunctional cantilever beam, which simplifies the complexity of the device.

3. Experimental Design

Firstly, this paper explores the nonlinear effect and designs two fixtures; one is an ordinary linear fixture, the other is a curved fixture. The design will test the respective characteristics of the two fixtures in an experiment for comparative analysis. First of all, based on these two fixtures, stoppers will be added at the middle and the end of the ordinary linear fixtures to explore the nonlinear effects of piezoelectric cantilever beams under three different structures. When the cantilever beam is deformed by external excitation, the cantilever beam will come into contact with the fixture, the effective length of the beam will be shortened, and the rigidity of the system will increase, thereby introducing nonlinear effects.

As shown in Figure 5, it shows the mechanical structure of the cantilever beam with stoppers at the middle and end of the cantilever on the basis of ordinary linear clamps and the use of curved clamps. The two cases of the ordinary linear fixture setting stoppers are shown in Figure 5a,b. The stoppers are, respectively, set at the middle and the end of the cantilever beam. When the cantilever beam is deformed due to vibration, the cantilever beam contacts the stopper; at this time, the effective length of the cantilever beam will change greatly, but this change is fixed. Figure 5c shows a curved fixture, which makes the effective length of the cantilever beam have different changes in one working cycle. Therefore, the nonlinear effect curves of the three structures are also significantly different.

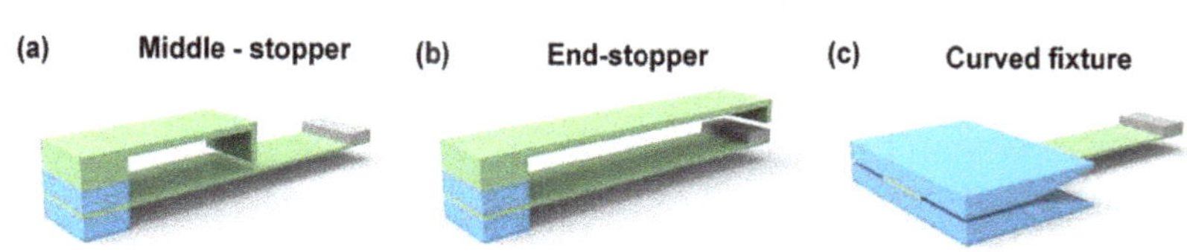

Figure 5. The mechanical structure of three non-linear test structures.

Additionally, this paper has carried on a certain theoretical derivation. In the general linear piezoelectric vibration power generation theoretical model, M_{eq} is the equivalent mass of the piezoelectric power generation module, ξ is the damping coefficient, K_{eq} is the equivalent stiffness of the piezoelectric power generation module, $u(t)$ is the vibration of the cantilever piezoelectric vibrator displacement, ∂ is the electromechanical coupling coefficient of the piezoelectric power generation module, and R_L is a purely resistive load. When the external excitation $F(t)$ affects, the dynamic equation of the equivalent mass block can be obtained according to Newton's second law, and the lumped parameter equivalent model of the piezoelectric power generation module shown in the following Equation (1) can be obtained according to Kirchhoff's law:

$$\begin{cases} M_{eq}\ddot{u}(t) + \xi\dot{u}(t) + K_{eq}u(t) + \partial V_0(t) = F(t) \\ \partial\dot{u}(t) = C_p\dot{V}_0(t) + V_0(t)/R_L \end{cases} \tag{1}$$

where $V_0(t)$ is the output voltage of the piezoelectric power generation module, and $\partial V_0(t)$ reflects the reaction force of inverse piezoelectric effect on the cantilever piezoelectric vibrator.

In the three fixtures shown in Figure 5, a theoretical model is established for two structures of the nonlinear piezoelectric power generation module with a stopper in Figure 5a,b. The damping coefficient, equivalent stiffness, and electromechanical coupling coefficient are expressed as a function of vibration displacement $\xi(u)$, $K_{eq}(u)$, and $\Theta(u)$, in order to obtain the general form of the system model of the nonlinear piezoelectric power generation module, as shown in the following Equation (2):

$$\begin{cases} M_{eq}\ddot{u}(t) + \xi(u)\dot{u}(t) + K_{eq}(u)u(t) + \Theta(u)Q(t)/C_p = F(t) \\ \Theta(u)u(t) = C_pQ(t)R_L + Q(t) \\ \xi(u) = a_1 + a_2|\dot{u}| \\ K_{eq} = b_1 + b_2u^2 \\ \Theta(u) = d_1 + d_2\sqrt{u} \end{cases} \tag{2}$$

where $Q(t)$ is the charge generated at both ends of the electrode; $\xi(u)$ is the square damping, where α_1 is the linear damping coefficient, and α_2 is the nonlinear damping coefficient; $K_{eq}(u)$ is the nonlinear stiffness, where b_1 is the linear stiffness coefficient, b_2 is the nonlinear stiffness coefficient; and $\Theta(u)$ is nonlinear electromechanical coupling coefficient, where d_1 is the linear electromechanical coupling coefficient, and d_2 is the nonlinear electromechanical coupling coefficient.

After integration and dimensionless processing, the vibration displacement, output charge, and time are standardized through the following Equation (3), and the theoretical general model of the nonlinear piezoelectric power generation module is obtained, as shown in Equation (4):

$$\begin{cases} u(t) = c_zz(\tau) \\ Q(t) = c_qq(\tau) \\ t = \tau\sqrt{m/b_1} \end{cases} \tag{3}$$

where c_z and c_q are standardization coefficients in units of m and C, respectively. τ is the standardization time.

$$\begin{cases} \ddot{z} + (\frac{a_1}{\sqrt{mb_1}}\dot{z} + \frac{c_za_2}{m}|\dot{z}|\dot{z}) + (z + \frac{b_2c_z{}^2}{b1}z^3) + \frac{c_q}{b_1c_zC_p}(d_1 + d_2\sqrt{c_z|z|})q \\ = F(\tau\sqrt{m/b_1})/(b_1c_z) \\ C_pR_L\dot{q}\sqrt{\frac{b_1}{m}} - \frac{c_z}{c_q}(d_1 + d_2\sqrt{c_z|z|})z + q = 0 \end{cases} \tag{4}$$

Simplified:

$$\begin{cases} \ddot{z} + (2\mu\dot{z} + \eta|\dot{z}|\dot{z}) + (z + \varphi z^3) + \varepsilon(\theta + \beta\sqrt{|z|})q \\ = F(\tau\sqrt{m/b_1})/(b_1c_z) \\ \rho\dot{q} - (\theta + \beta\sqrt{|z|})z + q = 0 \end{cases} \tag{5}$$

Making $2\mu = \alpha_1/\sqrt{mb_1}$, $\eta = c_z\alpha_2/m$, $\varphi = b_2c_z^2/b_1$, $\varepsilon = c_q^2/(b_1c_z^2C_p)$, $\theta = d_1c_z/c_q$, $\beta = d_2c_z\sqrt{c_z}/c_q$, and $\rho = C_pR_L\sqrt{b_1/m}$. z is the dimensionless vibration displacement, μ is the dimensionless linear damping coefficient, η is the dimensionless nonlinear damping coefficient, φ is the dimensionless nonlinear stiffness coefficient, θ is the dimensionless linear electromechanical coupling coefficient, β is the dimensionless nonlinear electromechanical coupling coefficient, q is the dimensionless output charge, and ρ is the dimensionless pure resistance load.

When the input excitation is $F(t) = F_0 sin(\Omega' t)$, the normalized formula on the right side of Equation (5) is as follows:

$$\hat{F}(\tau) = F_0/(b_1 c_z) sin(\Omega' \sqrt{m/b_1 \tau}) = \gamma sin(\omega\tau) \tag{6}$$

where $\gamma = F_0/(b_1 c_z)$ and $\omega = \sqrt{m/b_1}\Omega'$. Equation (5) is further transformed into the following:

$$\begin{cases} \ddot{z} + (2\mu\dot{z} + \eta|\dot{z}|\dot{z}) + (z + \varphi z^3) + \varepsilon(\theta + \beta\sqrt{|z|})q \\ = \gamma sin(\omega\tau) \\ \rho\dot{q} - (\theta + \beta\sqrt{|z|})z + q = 0 \end{cases} \tag{7}$$

Then, the average power in a period T is as follows:

$$\widetilde{P}_{avg} = \frac{1}{T}\int_0^T \rho\dot{q}(\tau)^2 \, d\tau \tag{8}$$

which, when converted to dimensional average power, is as follows:

$$P_{avg} = \frac{c_q^2}{C_q}\sqrt{\frac{b_1}{m}}\widetilde{P}_{avg} \tag{9}$$

Figure 5c shows the designed nonlinear piezoelectric power generation module of the bending fixture. The elastic restoring force $F_r = K_{eq}(u)u(t)$ is added to the established theoretical general model of the nonlinear piezoelectric power generation module to obtain the theoretical model of the nonlinear piezoelectric power generation module of the bending fixture, as shown in Equation (6).

$$\begin{cases} M_{eq}\ddot{u}(t) + \xi\dot{u}(t) + F_r + \partial V_0(t) = F(t) \\ \partial\dot{u}(t) = R_L C_p \dot{V}_0(t) + V_0(t) \\ F_r = f_1 u(t) + f_2 u(t)^3 \\ k = dF/du(t) = f_1 + 3f_2 u(t)^2 \end{cases} \tag{10}$$

where k is the nonlinear stiffness, where f_1 is the linear stiffness coefficient, and f_2 is the nonlinear stiffness coefficient.

Secondly, this paper also designs the experiment of the coupling characteristics among the three power generation modules, and explores the respective power generation characteristics of the three power generation modules and discusses the effect that the coupling can achieve.

4. Testing Results

Firstly, this paper explores the influence of the ordinary linear fixture and the curved fixture on the performance of piezoelectric power generation structures. As shown in Figure 6, when the external load is 10 kΩ and the excitation amplitude increases from 1.5 to 4 m/s^2, the non-linear effect exhibited by the curved fixture makes the frequency band continue to expand, which shows the good non-linear effect of the curved fixture. It can be seen that the bandwidth of the curved fixture is 25 to 33 Hz under 3 m/s^2, while the bandwidth of the ordinary linear fixture is 22 to 27 Hz. The curved fixture bandwidth under 3.5 m/s^2 is 26 to 35 Hz, and the bandwidth of the ordinary linear fixture is 22 to 27 Hz. At 3 and 3.5 m/s^2, the curved fixture has an expanded bandwidth compared to the ordinary linear fixture. Compared with the ordinary linear fixture, the curved fixture's bandwidth increased by 60 and 80%, respectively. At this time, the power of the curved fixture is 112 and 120% of the ordinary linear fixture.

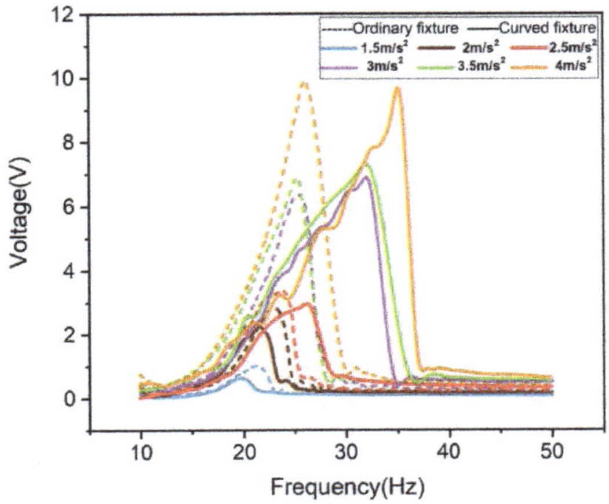

Figure 6. Ordinary fixture and curved fixture test comparison.

Secondly, based on the ordinary linear fixture, the experiment set stoppers at the middle and the end of the cantilever, and the nonlinear effects of the cantilever caused by these two structures and using the curved fixture are tested in Figure 7. It can be seen in the figure that although the non-linearity of placing the stopper at the end may not be obvious due to the small acceleration, it can be seen that the nonlinear effect of placing the stopper at the middle and using the curved fixture is more obvious. Obviously, the bandwidth and performance of using the curved fixture have been greatly improved compared to the ordinary linear fixture of setting stoppers. Using the curved fixture, the performance of the device is improved by nearly two times.

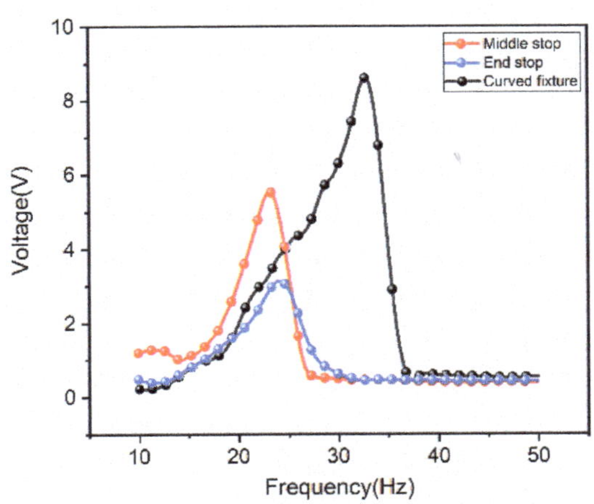

Figure 7. Three structural non-linear tests.

After exploring the advantages of the curved fixture, the experiment adds the electrostatic power generation module on this basis. The result is shown in Figure 8. After adding the electrostatic power generation module, because of the structure design of the electrostatic power generation module, it also shows a nonlinear effect. The half-power bandwidth of the piezoelectric cantilever beam is 14 to 19 Hz, and the half-power bandwidth of the electrostatic power generation module is 14 to 19.5 Hz. At this time, the electrostatic external load is 10 kΩ and the peak power is 3.6 mW, while the piezoelectric external load is 2 kΩ and the peak power is 2.2 mW. It can be seen that although the addition of the electrostatic power generation module does not expand the frequency bandwidth, it improves the performance of the generator. This means that under the same space utilization, the performance is improved by 60%.

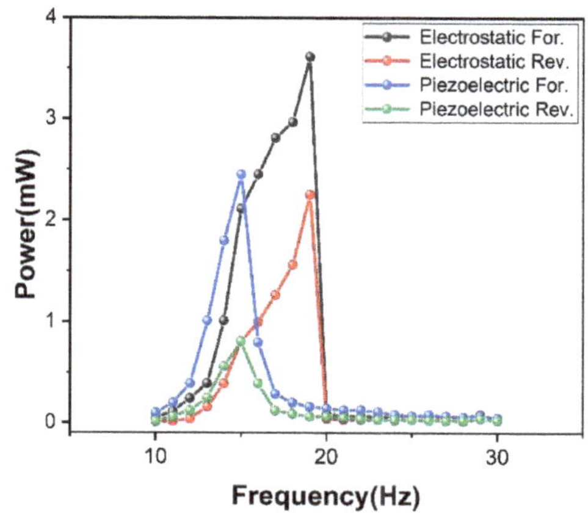

Figure 8. Non-linear curve with electrostatic structure added. For. is forward sweep and Rev. is reverse sweep.

In addition, this paper also explores the specific performance of the electromagnetic power generation module. After testing, the contribution of the electromagnetic module is 0.5 V of the voltage. The voltage test is shown in Figure 9. Although the electromagnetic power generation module has no nonlinear effect and does not broaden the frequency band of the device, its high current characteristics can improve the output current of the device.

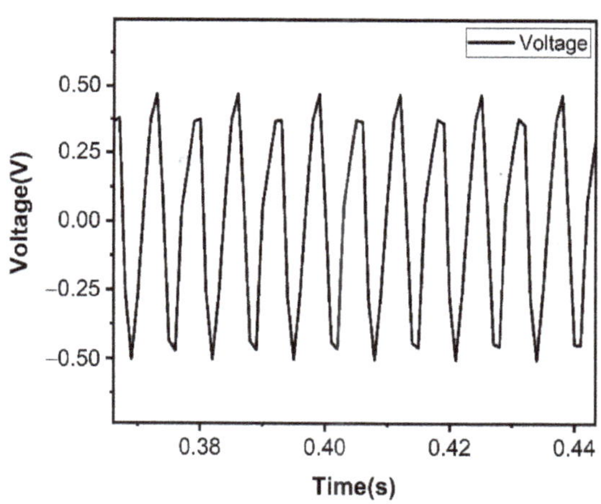

Figure 9. The voltage test of the electromagnetic power generation module.

To explore the influence of the high current characteristics of the electromagnetic module on the output current of the device, a design experiment is designed to test the current of the three modules and the current of the entire device. As shown in Figure 10, the electromagnetic high-current characteristics are well preserved, which greatly improves the current output of the device. The electrostatic module current can reach 28 μA, the piezoelectric module current can reach 7 μA, and the electromagnetic module current can reach 764 μA. Although the hybrid current of the three modules has a loss, it can still be seen that the electromagnetic module has increased the device current. The hybrid current can reach 653 μA.

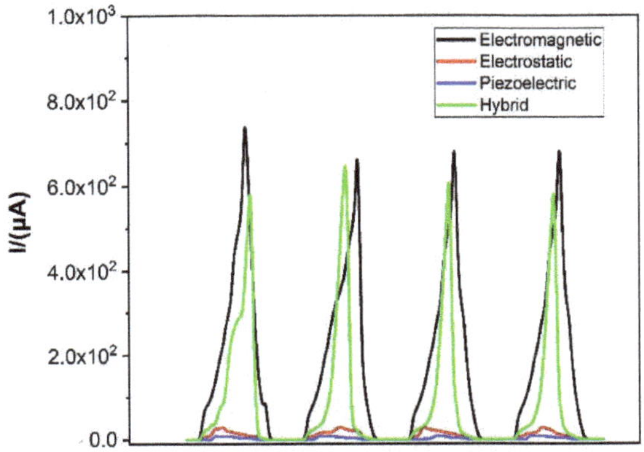

Figure 10. Three kinds of module current comparison and hybrid current.

In this paper, three kinds of power generation modules are integrated and encapsulated in the 3D printing shell together with the energy conditioning circuit. Figure 11 shows the appearance of the device after packaging and the energy management strategy used in this paper. In the paper, the impedance of PE is 2 kΩ, the impedance of ES is 10 kΩ, and the impedance of EM is only 170 Ω; the electrical impedance of the EM is significantly different from the PE and ES part. To overcome the problem, an additional impedance matching circuit needs to be incorporated in PE and ES after the AC-DC conversion, respectively. The impedance matching circuit is capable of adjusting the impedance of PE and ES to the same level of EM. After that, the DC signals from three mechanisms are superposed together for an improved performance.

This paper integrates the three power generation modules and the energy conditioning circuit in a small-sized package shell. Under ensuring its good performance, the size of the device is reduced as much as possible so that the device can adapt to more application scenes. According to the calculation, the circuit power consumption of the current system is calculated up to 3 mW, and the overall converted energy before the power conditioning circuit is about 6.5 mW. The conversion efficiency of mechanical energy to electrical energy ratio is calculated as 0.12/1.6 = 0.075, 7.5%.

Three kinds of power generation modules are effectively integrated into the confined space. During the working process of the device, the electromagnetic power generation module actively participates in the contribution in the low-frequency domain, and its high current characteristic is fully utilized. In the high-frequency domain, the electrostatic power generation module and the piezoelectric power generation module actively participate in the contribution, and the high-voltage characteristics are fully exerted. Therefore, the device has a wider working frequency domain and achieves a higher energy conversion efficiency in multiple vibration frequency bands. In the same time domain, the multifunctional cantilever beam structure means that the output of the piezoelectric, electrostatic, and electromagnetic power generation modules is basically in the same phase. The integrated output loss is small, and the characteristics of high voltage and high current are realized.

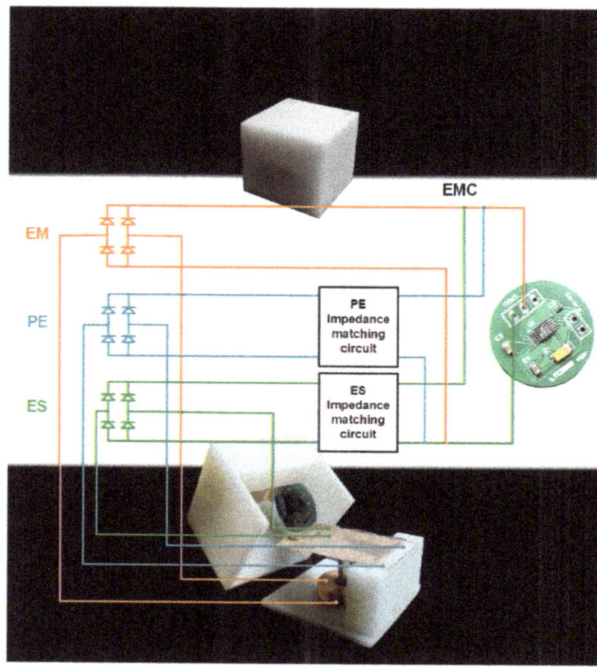

Figure 11. The physical map of the device and the energy conditioning circuit.

Finally, there is an energy conditioning circuit at the rear end of the device. The output of the three power generation modules is adjusted by the energy conditioning circuit and the AC is converted into the DC that most sensors can directly use. The conversion effect is shown in Figure 12. After the conditioning of the energy conditioning circuit, it can finally output a 3.3 V DC voltage.

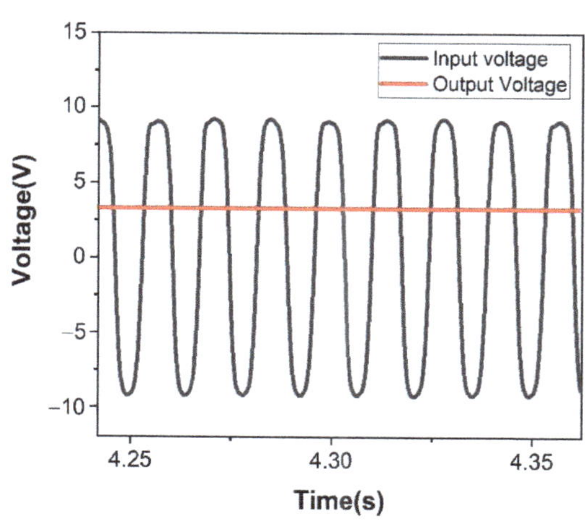

Figure 12. The conversion effect of the energy conditioning circuit.

5. Conclusions

This paper proposes a hybrid nonlinear generator with the following three power generation modules: piezoelectric, electrostatic, and electromagnetic. The influence of the ordinary linear fixture and the curved fixture on the performance of the generator has been explored. It has been verified that compared with the ordinary linear fixture, the curved fixture has excellent bandwidth expansion capabilities and can improve the performance by 25% compared with the ordinary stopper structure. While exploring the nonlinear effect of the piezoelectric cantilever beam, we have integrated the electrostatic power generation module and the electromagnetic power generation module into the piezoelectric power generation module. After adding the electrostatic power generation module, the half-power bandwidth of the piezoelectric cantilever beam has been increased from 16 to 20 Hz. The peak power of 3.6 mW has been achieved. The half-power bandwidth of the electrostatic structure has been increased from 14.5 to 19.5 Hz, and the peak power is 2.2 mW. Although the addition of electrostatic energy generation module does not expand the frequency bandwidth, it improves the performance of the generator. This means that under the same space utilization, the performance is improved by 60%. In addition, the electromagnetic energy generation module is introduced into the cantilever beam structure, which also contributes to the device. Finally, we have optimized the structure and performance of the hybrid generator and have achieved the improvement of the bandwidth and performance. Moreover, an energy conditioning circuit at the rear end of the device is added. Through the energy conditioning circuit, the device can directly output a DC voltage to supply power to most of the sensing equipment.

In general, the device designed in this paper has the advantages of a larger bandwidth, better performance, and smaller size, and can be used in many vibration environments in limited spaces. Additionally, the device can overcome the shortcomings of the single energy harvesting device, such as the low current of piezoelectric and electrostatic power generation, and the low voltage of electromagnetic power generation, etc., the output performance is better than the single energy harvesting device. Although the hybrid energy harvesting device has these advantages, it also has some problems, such as difficulty in coupling multiple power generation units, large size, and high cost. Therefore, further improving the coupling performance of multiple units of the hybrid energy harvesting device, designing new structures to improve integration and reduce the device size, and reducing costs are things that need to be considered.

Author Contributions: Conceptualization, Y.C. and F.L.; methodology, Y.C.; software, H.Q.; validation, Y.Y. and K.T.; investigation, J.W.; resources, W.S. and J.M.; data curation, Y.C.; writing—review and editing, F.L., J.W., Y.Y. and K.T.; funding acquisition, W.S., H.Q. and J.M. All authors have read and agreed to the published version of the manuscript.

Funding: This research is supported by National Natural Science Foundation of China Grant No. 51705429, the Fundamental Research Funds for the Central Universities and Research & Development Institute of Northwestern Polytechnical University, Shenzhen 518057, China, grant number Shenzhen Research Plan (JCYJ20180306171637410).

Data Availability Statement: The data that support the findings of this study are available from the first author upon reasonable request.

Conflicts of Interest: The authors declare no conflict of interest.

References

1. Pirisi, A.; Mussetta, M.; Gruosso, G.; Zich, R.E. Bio-inspired optimization techniques for wireless energy transfer. In Proceedings of the International Conference on Electromagnetics in Advanced Applications (ICEAA), Sydney, Australia, 20–24 September 2010; pp. 731–732.
2. Liu, H.C.; Zhong, J.W.; Lee, C.K.; Lee, S.-W.; Lin, L.W. A comprehensive review on piezoelectric energy harvesting technology: Materials, mechanisms, and applications. *Appl. Phys. Rev.* **2018**, *5*, 041306. [CrossRef]
3. Li, Y.J.; Tao, K.; George, B.; Tan, Z.C. Harvesting Vibration Energy: Technologies and Challenges. *IEEE Ind. Electron. Mag.* **2020**, *15*, 30–39. [CrossRef]
4. Yang, J.; Zhong, J.M.; Chang, H.L. A Closed-Loop Mode-Localized Accelerometer. *IEEE/ASME J. Microelectromech. Syst.* **2018**, *27*, 210–217. [CrossRef]
5. Wang, J.L.; Tang, L.H.; Zhao, L.Y.; Hu, G.B.; Song, R.J.; Xu, K. Equivalent circuit representation of a vortex-induced vibration-based energy harvester using a semi-empirical lumped parameter approach. *Int. J. Energy Res.* **2020**, *44*, 4516–4528. [CrossRef]

6. Wu, Z.X.; Yang, X.; Wu, J. Conductive Hydrogel- and Organohydrogel-Based Stretchable Sensors. *ACS Appl. Mater. Interfaces* **2021**, *13*, 2128–2144. [CrossRef] [PubMed]
7. Zhao, L.C.; Zou, H.X.; Yan, G.; Liu, F.R.; Tan, T.; Wei, K.X.; Zhang, W.M. Magnetic coupling and flextensional amplification mechanisms for high-robustness ambient wind energy harvesting. *Energy Convers. Manag.* **2019**, *201*, 112166. [CrossRef]
8. Wu, Z.X.; Ding, H.J.; Tao, K.; Wei, Y.M.; Gui, X.C.; Shi, W.X.; Xie, X.; Wu, J. Ultrasensitive, Stretchable, and Fast-Response Temperature Sensors Based on Hydrogel Films for Wearable Applications. *ACS Appl. Mater. Interfaces* **2021**, *13*, 21854–21864. [CrossRef]
9. Cao, H.L.; Zhang, Y.J.; Han, Z.Q.; Shao, X.L.; Gao, J.Y.; Huan, K.; Shi, Y.B.; Tang, J.; Shen, C.; Liu, J. Pole-zero temperature compensation circuit design and experiment for dual-mass MEMS gyroscope bandwidth expansion. *IEEE-ASME Trans. Mechatron.* **2019**, *24*, 677–688. [CrossRef]
10. Tao, K.; Yi, H.P.; Tang, L.H.; Wu, J.; Wang, P.H.; Wang, N.; Hu, L.X.; Fu, Y.Q.; Miao, J.M.; Chang, H.L. Piezoelectric ZnO thin films for 2DOF MEMS vibrational energy harvesting. *Surf. Coat. Technol.* **2019**, *359*, 289–295. [CrossRef]
11. Shearwood, C.; Yates, R. Development of an electromagnetic microgenerator. *Electron. Lett.* **1997**, *33*, 1883–1884. [CrossRef]
12. Meninger, S.; Mur-Miranda, J.O.; Amirtharajah, R.; Chandrakasan, A.; Lang, J.H. Vibration-to-electric energy conversion. *IEEE Trans. Very Large Scale Integr. Syst.* **2001**, *9*, 64–74. [CrossRef]
13. Fan, K.Q.; Liu, J.; Wei, D.M.; Zhang, D.X.; Zhang, Y.; Tao, K. A cantilever-plucked and vibration-driven rotational energy harvester with high electric outputs. *Energy Convers. Manag.* **2021**, *244*, 114504. [CrossRef]
14. Wang, J.L.; Sun, S.K.; Tang, L.H.; Hu, G.B.; Liang, J.L. On the use of metasurface for Vortex-Induced vibration suppression or energy harvesting. *Energy Convers. Manag.* **2021**, *235*, 113991. [CrossRef]
15. Fang, H.B.; Liu, J.Q.; Xu, Z.Y.; Dong, L.; Wang, L.; Chen, D.; Cai, B.C.; Liu, Y. Fabrication and performance of MEMS-based piezoelectric power generator for vibration energy harvesting. *Microelectron. J.* **2006**, *37*, 1280–1284. [CrossRef]
16. Wang, L.; Yuan, F.G. Vibration energy harvesting by magnetostrictive material. *Smart Mater. Struct.* **2008**, *17*, 045009. [CrossRef]
17. Tao, K.; Wu, J.; Tang, L.H.; Xia, X.; Lye, S.W.; Miao, J.M.; Hu, X. A novel two-degree-of-freedom MEMS electromagnetic vibration energy harvester. *J. Micromech. Microeng.* **2016**, *26*, 035020. [CrossRef]
18. Wang, J.L.; Hu, G.B.; Su, Z.; Li, G.P.; Zhao, W.; Tang, L.H.; Zhao, L.Y. A cross-coupled dual-beam for multi-directional energy harvesting from vortex induced vibrations. *Smart Mater. Struct.* **2019**, *28*, 12LT02. [CrossRef]
19. Shan, X.B.; Li, H.L.; Yang, Y.C.; Feng, J.; Wang, Y.C.; Xie, T. Enhancing the performance of an underwater piezoelectric energy harvester based on flow-induced vibration. *Energy* **2019**, *172*, 134–140. [CrossRef]
20. Wu, Y.; Qiu, J.; Zhou, S.; Ji, H.; Chen, Y.; Li, S. A piezoelectric spring pendulum oscillator used for multidirectional and ultra-low frequency vibration energy harvesting. *Appl. Energy* **2018**, *231*, 600–614. [CrossRef]
21. Chen, J.; Zhu, G.; Yang, W.; Jing, Q.; Bai, P.; Yang, Y.; Hou, T.C.; Wang, Z.L. Harmonic-resonator-based triboelectric nanogenerator as a sustainable power source and a self-powered active vibration sensor. *Adv. Mater.* **2013**, *25*, 6094–6099. [CrossRef]
22. Lu, W.L.; Hwang, Y.M. Analysis of a vibration-induced micro-generator with a helical micro-spring and induction coil. *Microelectron. Reliab.* **2012**, *52*, 262–270. [CrossRef]
23. Si, H.; Dong, J.L.; Chen, L.; Sun, L.Z.; Zhang, X.D.; Gao, M.T. Study of the ambient vibration energy harvesting based on piezoelectric effect. *Int. J. Nanosci.* **2015**, *14*, 1460017. [CrossRef]
24. Huang, J.K.; O'Handley, R.C.; Bono, D. New high-sensitivity hybrid magnetostrictive/electroactive magnetic field sensors. In Proceedings of the Smart Structures and Materials 2003: Smart Sensor Technology and Measurement Systems, San Diego, CA, USA, 3–5 March 2003; Volume 5050, pp. 229–237.
25. Karami, M.A.; Inman, D.J. Nonlinear hybrid energy harvesting utilizing a piezo -magneto-elastic spring. In Proceedings of the Active and Passive Smart Structures and Integrated Systems 2010, San Diego, CA, USA, 8–11 March 2010; Volume 7643, pp. 379–380.
26. Challa, V.R.; Prasad, M.G.; Fisher, F.T. A coupled piezoelectric-electromagnetic energy harvesting technique for achieving increased power output through damping matching. *Smart Mater. Struct.* **2009**, *18*, 7566–7579. [CrossRef]
27. Tao, K.; Chen, Z.S.; Yi, H.P.; Zhang, R.; Shen, Q.; Wu, J.; Tang, L.; Fan, K.; Fu, Y.; Miao, J.; et al. Hierarchical Honeycomb-Structured Electret/Triboelectric Nanogenerator for Biomechanical and Morphing Wing Energy Harvesting. *Nano-Micro Lett.* **2021**, *13*, 1–6. [CrossRef] [PubMed]
28. Naruse, Y.; Matsubara, N.; Mabuchi, K.; Izumi, M.; Suzuki, S. Electrostatic micro power generation from low-frequency vibration such as human motion. *J. Micromech. Microeng.* **2009**, *19*, 094002. [CrossRef]
29. Miki, D.; Honzumi, M.; Suzuki, Y.; Kasagi, N. Large-amplitude MEMS electret generator with nonlinear spring. In Proceedings of the 2010 IEEE 23rd International Conference on Micro Electro Mechanical Systems (MEMS), Hong Kong, China, 24–28 January 2010; pp. 176–179.
30. Erturun, U.; Eisape, A.; West, J.E. Design and analysis of a vibration energy harvester using push-pull electrostatic conversion. *Smart Mater. Struct.* **2020**, *29*, 105018. [CrossRef]
31. Erturun, U.; Eisape, A.A.; Kang, S.H.; West, J.E. Energy harvester using piezoelectric nanogenerator and electrostatic generator. *Appl. Phys. Lett.* **2021**, *118*, 063902. [CrossRef]
32. Kulkarni, S.; Roy, S.; O'Donnell, T.; Beeby, S.; Tudor, J. Vibration based electromagnetic micropower generator on silicon. *J. Appl. Phys.* **2006**, *99*, 08P511. [CrossRef]

33. Kulah, H.; Najafi, K. Energy scavenging from low-frequency vibrations by using frequency up-conversion for wireless sensor applications. *IEEE Sens. J.* **2008**, *8*, 261–268. [CrossRef]
34. Wang, P.H.; Liu, H.T.; Yang, Z.Q.; Dai, X.H.; Zhao, X.L. Sandwiched Electromagnetic Vibration Energy Harvester Based on MEMS Technology. *Nanotechnol. Precis. Eng.* **2010**, *6*, 510–515.
35. Paul, K.; Mallick, D.; Roy, S. Performance improvement of MEMS Electromagnetic Vibration Energy Harvester using optimized patterns of micromagnet array. *IEEE Magn. Lett.* **2021**, *12*, 2101805. [CrossRef]
36. Bai, Y.; Tofel, P.; Hadas, Z.; Smilek, J.; Losak, P.; Skarvada, P.; Macku, R. Investigation of a cantilever structured piezoelectric energy harvester used for wearable devices with random vibration input. *Mech. Syst. Signal Process.* **2018**, *106*, 303–318. [CrossRef]
37. Shahruz, S.M. Design of mechanical band-pass filters for energy scavenging. *J. Sound Vib.* **2006**, *292*, 987–998. [CrossRef]
38. Xue, H.A.; Hu, Y.T.; Wang, Q.M. Broadband piezoelectric energy harvesting devices using multiple bimorphs with different operating frequencies. *IEEE Trans. Ultrason. Ferroelectr. Freq. Control* **2008**, *55*, 2104–2108. [PubMed]
39. Leland, E.S.; Wright, P.K. Resonance tuning of piezoelectric vibration energy scavenging generators using compressive axial preload. *Smart Mater. Struct.* **2006**, *15*, 1413–1420. [CrossRef]
40. Li, H.T.; Yang, Z.; Zu, J.; Qin, W.Y. Numerical and experimental study of a compressive-mode energy harvester under random excitations. *Smart Mater. Struct.* **2017**, *26*, 035064. [CrossRef]
41. Khazaee, M.; Rezaniakolaie, A.; Rosendahl, L. A broadband macro-fiber-composite piezoelectric energy harvester for higher energy conversion from practical wideband vibrations. *Nano Energy* **2020**, *76*, 104978. [CrossRef]
42. Liu, W.Q.; Liu, C.Z.; Ren, B.Y.; Zhu, Q.; Hu, G.D.; Yang, W.Q. Bandwidth increasing mechanism by introducing a curve fixture to the cantilever generator. *Appl. Phys. Lett.* **2016**, *109*, 043905. [CrossRef]
43. Tao, K.; Yi, H.P.; Yang, Y.; Tang, L.H.; Yang, Z.S.; Wu, J.; Chang, H.L.; Yuan, W.Z. Miura-origami-inspired electret/triboelectric power generator for wearable energy harvesting with water-proof capability. *Microsyst. Nanoeng.* **2020**, *6*, 1–11. [CrossRef]
44. Zhang, X.Q.; Pondrom, P.; Sessler, G.M.; Ma, X.C. Ferroelectret nanogenerator with large transverse piezoelectric activity. *Nano Energy* **2018**, *50*, 52–61. [CrossRef]
45. Tao, K.; Tang, L.H.; Wu, J.; Lye, S.W.; Chang, H.L.; Miao, J.M. Investigation of multimodal electret-based MEMS energy harvester with impact-induced nonlinearity. *J. Microelectromech. Syst.* **2018**, *27*, 276–288. [CrossRef]
46. Zhang, Y.L.; Wang, T.Y.; Luo, A.X.; Hu, Y.S.; Li, X.; Wang, F. Micro electrostatic energy harvester with both broad bandwidth and high normalized power density. *Appl. Energy* **2018**, *212*, 362–371. [CrossRef]
47. Tao, K.; Yi, H.P.; Yang, Y.; Chang, H.L.; Wu, J.; Tang, L.H.; Yang, Z.S.; Wang, N.; Hu, L.X.; Fu, Y.Q.; et al. Origami-inspired electret-based triboelectric generator for biomechanical and ocean wave energy harvesting. *Nano Energy* **2020**, *67*, 104197. [CrossRef]
48. Tao, K.; Lye, S.W.; Tang, L.H.; Miao, J.M.; Hu, X. Out-of-plane electret-based MEMS energy harvester with the combined nonlinear effect from electrostatic force and a mechanical elastic stopper. *J. Micromech. Microeng.* **2015**, *10*, 104014. [CrossRef]

micromachines

MDPI

Article

Analysis of the Influence of Ferromagnetic Material on the Output Characteristics of Halbach Array Energy-Harvesting Structure

Xiangyong Zhang [1], Haipeng Liu [2,*], Yunli He [3], Tingrui Peng [1], Bin Su [1] and Huiyuan Guan [1]

[1] School of Mechatronical Engineering, Beijing Institute of Technology, Beijing 100081, China; zhangxiangyong177@163.com (X.Z.); 3120200131@bit.edu.cn (T.P.); subinbaiyang@163.com (B.S.); ghy907553326@163.com (H.G.)
[2] State Key Laboratory of Explosion Science and Technology, Beijing Institute of Technology, Beijing 100081, China
[3] Department of Mechanical Engineering, Shandong Huayu University of Technology, Dezhou 253034, China; heyunli1986@163.com
* Correspondence: lhp@bit.edu.cn

Abstract: Due to the particular arrangement of permanent magnets, a Halbach array has an significant effect of magnetism and magnetic self-shielding. It can stretch the magnetic lines on one side of the magnetic field to obtain an ideal sinusoidal unilateral magnetic field. It has a wide application range in the field of energy harvesting. In practical applications, magnetic induction intensity of each point in magnetic field is not only related to the induced current and conductor but also related to the permeability of the medium (also known as a magnetic medium) in the magnetic field. Permeability is the physical quantity that represents the magnetism of the magnetic medium, which indicates the resistance of magnetic flux or the ability of magnetic lines to be connected in the magnetic field after coil flows through current in space or in the core space. When the permeability is much greater than one, it is a ferromagnetic material. Adding a ferromagnetic material in a magnetic field can increase the magnetic induction intensity B. Iron sheet is a good magnetic material, and it is easy to magnetize to generate an additional magnetic field to strengthen the original magnetic field, and it is easy to obtain at low cost. In this paper, in order to explore the influence of ferromagnetic material on the magnetic field and energy harvesting efficiency of the Halbach array energy harvesting structure, iron sheets are installed on the periphery of the Halbach array rotor. Iron sheet has excellent magnetic permeability. Through simulation, angle between iron sheet and Halbach array, radian size of iron sheet itself and distance between iron sheet and Halbach array can all have different effects on the magnetic field of the Halbach array. It shows that adding iron sheets as a magnetic medium could indeed change the magnetic field distribution of the Halbach array and increase energy harvesting efficiency. In this paper, a Halbach array can be used to provide electrical power for passive wireless low-power devices.

Keywords: Halbach array; iron sheet; energy harvesting

Citation: Zhang, X.; Liu, H.; He, Y.; Peng, T.; Su, B.; Guan, H. Analysis of the Influence of Ferromagnetic Material on the Output Characteristics of Halbach Array Energy-Harvesting Structure. *Micromachines* **2021**, *12*, 1541. https://doi.org/10.3390/mi12121541

Academic Editor: Sang-Jae Kim

Received: 30 October 2021
Accepted: 9 December 2021
Published: 11 December 2021

1. Introduction

With the continuous development of micro devices, various wearable devices that can be used on the human body have been widely used. Zhihao Ren et al. based one on the smart cut process for fabrication of silicon-on-insulator (SOI) wafers and films made of wafer bonding technology, which can satisfy high-performance molecular sensors for next-generation applications beyond 5G [1] At the same time as the development of micro devices, energy harvest technology has also been developed. Xiang Lu et al. designed an energy trap based on the principle of piezoelectricity and the miniature deformable squama mechanics [2] by controlling the size of the signal input to each array to produce different shapes of deformation to fit the human body movement. The piezoelectric energy harvest voltage is large, but it has strict requirements on frequency and bandwidth.

In 1979, physicist Klaus Halbach discovered a particular permanent magnet array—the Halbach array [3–5]. Using a special magnet unit arrangement, field strength in the magnet unit direction increased. The goal is to use a small number of magnets to generate the strongest magnetic field. The study by Ni Y, Xu L, Jing L et al. showed that the Halbach array has a higher air gap magnetic flux density amplitude than conventional magnetic steel, which can significantly improve the energy harvesting efficiency [6–8].

It is crucial to study the distribution of the magnetic field in electromagnetic harvesting energy, and it is the prerequisite for applying a magnetic medium in a magnetic field to improve the magnetic field [9–12]. After ferromagnetic material that is magnetized in the magnetic field, an additional magnetic field will be generated. The superposition of additional magnetic field and original magnetic field makes the total magnetic field much stronger than the original magnetic field [13–15]. In this paper, iron sheets are used to explore the influence of ferromagnetic material on the magnetic field distribution of the Halbach array. As the position, size and angle of iron sheet are different, the final magnetic field distribution of the Halbach array will be different, and the energy harvesting efficiency will be different. Three sets of simulation comparisons are designed for verification. The results prove that when it is a ring-shaped Halbach array composed of three pole pairs and two permanent magnets per pole, and the angle between the iron sheet and Halbach array is 30°, the magnetic field distribution is the densest and the energy harvesting effect is the best. When the radian of the iron sheet is 40°, the amount of iron sheet is the lowest, while the energy harvesting effect is the best. When the distance between the iron sheet and Halbach array gets closer, the energy-harvesting effectiveness of the designed structure grows higher. Compared with the case of without iron sheet, the maximum energy harvesting voltage is increased by 1.8 times, and the maximum energy harvesting voltage is 1.8 V. The simulation results are verified through experiments, and when iron sheets are added, the energy harvesting effect is indeed greatly improved.

2. The Principle of Halbach Array and Ferromagnetic Magnetization

In order to make the air gap flux density of the Halbach array magnet rotor present a sinusoidal distribution, it is necessary to perform continuous sinusoidal magnetization of the permanent magnets to make up the Halbach array rotor, namely Halbach magnetization [16]. The ideal sinusoidal magnetization effect is better, but the process is complicated to achieve with the existing technology. Therefore, the permanent magnets are assembled and magnetized in segments. According to the calculus theory, as long as the permanent magnets are magnetized and discretely connected as much as possible on each pole pair of the Halbach array, an approximately sinusoidal air gap magnetic density waveform can be obtained [17,18]. In this paper, the Halbach array composed of three pole pairs and two permanent magnets per pole needs to be magnetized and arranged according to the following equations.

$$\beta = \frac{(p-1)\pi}{mp} \tag{1}$$

$$\theta = \frac{\pi}{mp} \tag{2}$$

where m is the number of pieces of the permanent magnet under each pole, θ is the angle of a single permanent magnet, β is the magnetizing angle of two adjacent permanent magnets in a single pole and p is the number of motor pole pairs.

The permanent magnet arrangement of Halbach array used in this paper are shown in Figure 1.

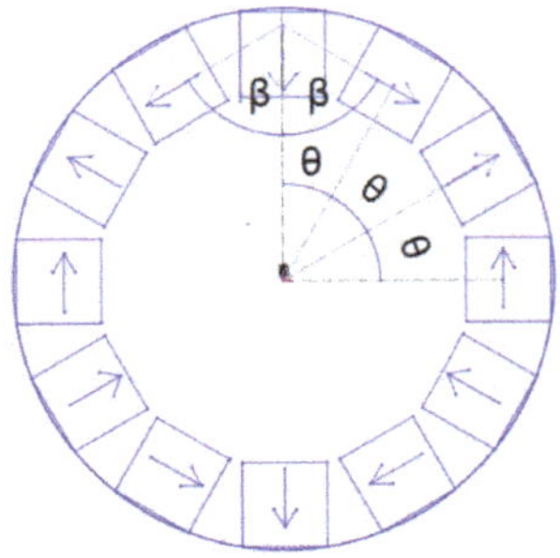

Figure 1. Permanent magnet arrangement of Halbach array.

For the Halbach array rotor with two permanent magnets per pole, each individual permanent magnet of the Halbach array is magnetized in a single direction, arranged in the radial and circumferential directions. The radially arranged permanent magnets mainly produce magnetic poles, and the tangentially arranged permanent magnets mainly guide and enhance the magnetic flux [19–21]. Because the unique structure of Halbach array can enhance the magnetic field strength on one side, its energy harvesting effect will also be better.

The Halbach array generates a sinusoidal magnetic field. When a ferromagnetic material is placed in the magnetic field, the ferromagnetic material is magnetized to produce an additional magnetic field. Magnetized body current J_m appears in the ferromagnetic material after magnetization; Magnetizing surface current K_m appears on the ferromagnetic surface.

$$J_m = \nabla \times M. \tag{3}$$

$$K_m = M \times e_n. \tag{4}$$

where M is the magnetization and e_n is the unit vector of the outer normal of ferromagnetic surface.

Using Stokes' theorem, from Equation (3) we can get

$$\oint_l M \cdot dl = \int_l (\nabla \times M) \cdot dS = I_m \tag{5}$$

where I_m is the magnetizing current.

The magnetic field in the presence of ferromagnetic material can be regarded as a synthetic magnetic field created by the original magnetic field and the magnetizing current. It can be obtained from the Ampere's loop law.

$$\oint_l B \cdot dl = \mu_0 \sum I_m + B_0 \tag{6}$$

where B_0 is the magnetic field strength of original magnetic field.

Therefore, it can be seen that adding ferromagnetic material in the magnetic field will indeed affect the original magnetic field. In this paper, the Halbach array produces a sinusoidal magnetic field, and the ferromagnetic material has different magnetization effects when placed in different positions.

Introduce a new field quantity—magnetic field strength H

$$H = \frac{B}{\mu_0} - M \tag{7}$$

where μ_0 is the vacuum permeability, and B is the synthetic magnetic field of original magnetic field and additional magnetic field generated by the ferromagnetic material.

For most ferromagnetic material, the magnetization is proportional to the magnetic field strength.

$$M = \chi_m H \tag{8}$$

where χ_m is the magnetic susceptibility, which is a pure one-dimensional number. Substituting Equation (8) into (7), we can get

$$B = \mu_0(1 + \chi_m)H \tag{9}$$

$$\text{Let } \mu = \mu_0(1 + \chi_m) \tag{10}$$

Get the combinatorial equation of ferromagnetic material

$$B = \mu H \tag{11}$$

where μ is the permeability of ferromagnetic material.

It can be seen from the above that when ferromagnetic material is added to the magnetic field of Halbach array rotor, the physical properties of ferromagnetic material itself will also affect the magnetization.

3. The Effect of Iron Sheet on Halbach Array

In this paper, the influence of iron sheet on the Halbach array from angle between iron sheet and Halbach array, radian size of iron sheet and distance between iron sheet and Halbach array is explored. The Halbach array has three pairs of poles; the six pieces of iron sheet are evenly arranged on the outside of the Halbach array rotor. The other parameters are shown in Table 1.

Table 1. Parameters used in the experiment.

Rectangular permanent magnet	5 mm × 5 mm × 5 mm
The radius of the Halbach array rotor	16 mm
The thickness of the iron sheet	1 mm
Permanent magnet residual magnetic flux density	0.84 T

The magnetic field distribution diagram of the Halbach array when there is no iron sheet is shown in Figure 2.

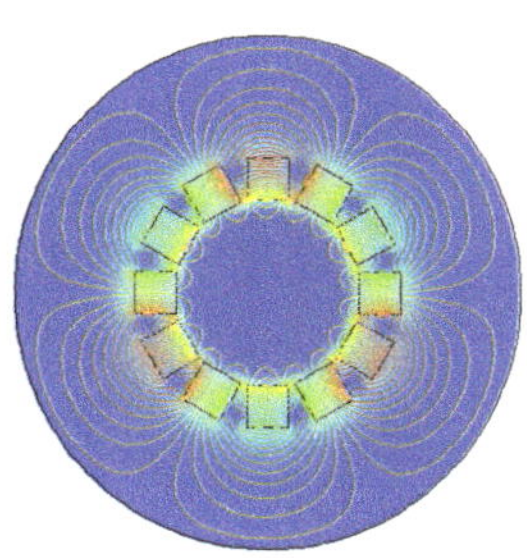

Figure 2. Halbach array magnetic field distribution diagram.

It can be seen from Figure 2 that the sinusoidal distribution of the magnetic field is better when there is no iron sheet. Using 6 groups of 100-turn coils, gap between the coil and the outer periphery of Halbach array rotor is 1 mm, and iron sheet is placed on the periphery of the coil with a gap of 0.5 mm from the coil. The radian of iron sheet is 40°for each piece, and the rotation rate is 10 r/s. The maximum energy harvesting voltage obtained by simulation is 1 V.

3.1. The Influence of the Angle between Halbach Array and Iron Sheet

In this paper, the coil is set to be fixed. The permanent magnet and the iron sheet are fixed as a whole to rotate together, and the positions of the permanent magnet and the iron sheet do not change during the rotation. The actual device is shown in Figure 3b

The Halbach array used in this paper is a three-pole pair. Thus, there are six rectangular permanent magnets arranged radially. There are six identical iron sheets and the distance between iron sheet is 3 mm from the Halbach array. In order to explore the relationship between the angle between iron sheet and the Halbach array, the angle between the iron sheet and the Halbach array is defined as the angle between the center of the iron sheet and the centerline of radially arranged permanent magnets with the magnetic poles, as shown in Figure 3a.

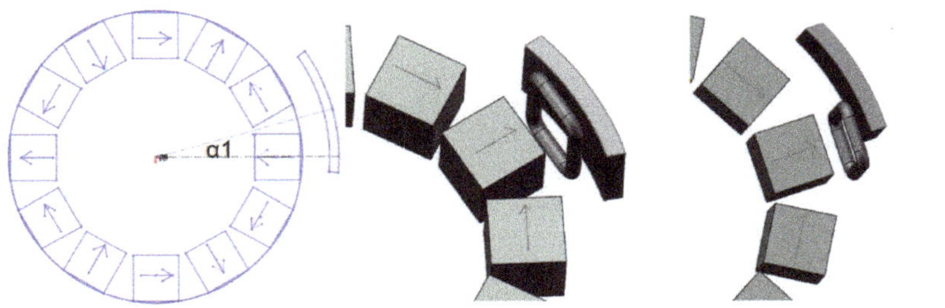

(**a**) Angle representation (**b**) No change in angle (**c**) Contrast of angle change

Figure 3. The angle between Halbach array and iron sheet.

It can be seen from the Figure 3b,c that the relative position of the iron sheet and the Halbach array rotor is different, and it can be seen from the Figure that the angle of 60° between iron sheet and the Halbach array forms a circle. Therefore, the magnetic field distribution of the Halbach array under different conditions from 0 to 60° is explored, as shown in Figure 4.

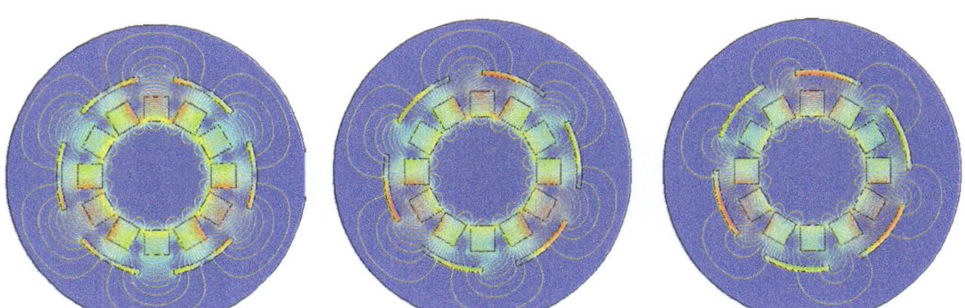

(**a**) The angle is 0° (**b**) The angle is 10° (**c**) The angle is 20°

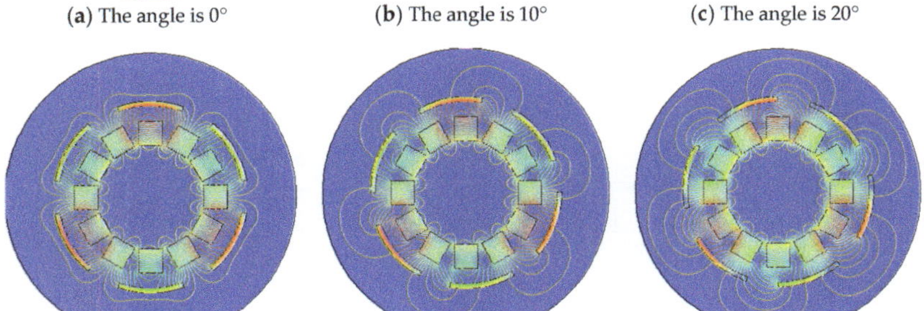

(**d**) The angle is 30° (**e**) The angle is 40° (**f**) The angle is 50°

Figure 4. The magnetic field distribution diagram of iron sheet and Halbach array at different angles.

It can be seen from Figure 4 above that the difference in the angle between iron sheet and the Halbach array will indeed cause the difference in the magnetic field distribution of the Halbach array. When the angle is 0°, the change in the magnetic field of the Halbach array by iron sheet is not apparent. Compared with no iron sheet, the magnetic lines are concentrated only in the gap between these two iron sheets. When the angle increases to 10°, the magnetic lines begin to stretch into iron sheet, and the magnetic field begins to gather in the gap between the iron sheet and the Halbach array. When the angle is 20°, the magnetization effect is more prominent, and the distribution of magnetic lines outside iron sheets is reduced. When the angle is 30°, magnetizing effect of iron sheet on the Halbach array reaches the maximum. The magnetic flux density in the gap is greatly enhanced, magnetic lines outside iron sheets are greatly reduced, and the magnetization effect is optimal. When the angle continues to increase, the magnetization effect begins to weaken. When the angle is 40°, the magnetic field distribution is the same as when it is 20°. When the angle is 50°, the magnetic field distribution is the same as when it is 10°. When the angle changes from 0 to 60°, its magnetization effect first increases and then decreases. To verify these conclusions, the energy harvesting simulation is carried out. The maximum voltage obtained under the same conditions as shown in Figure 5.

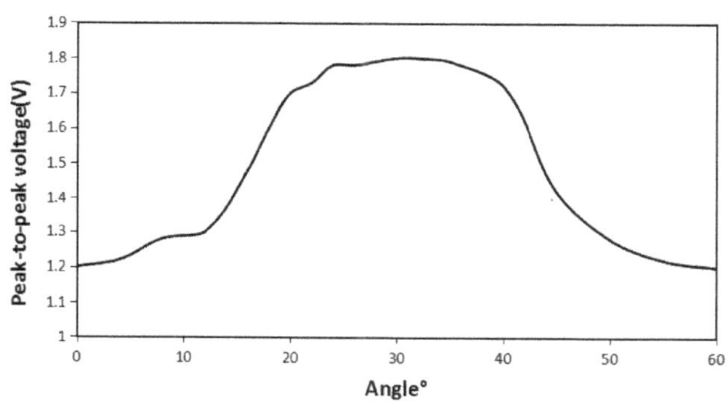

Figure 5. Peak-to-peak voltage of iron sheet and Halbach array rotor at different angles.

It can be seen from Figure 5 that when angle between the iron sheet and the Halbach array rotor is 0°, the energy harvesting effect is the worst. The peak-to-peak voltage is 1.2 V, which is an increase of 20% compared with case of the non-iron sheet. When angle increases, it is consistent with the magnetic field distribution result. Due to the concentration of iron sheet on magnetic field, the magnetic lines are stretched between the iron sheets and the Halbach array rotor, so its energy harvesting effect is greatly improved. When the angle is 30°, the peak-to-peak value of the harvesting energy voltage reaches the maximum of 1.8 V, which increases 80% compared to the case without iron sheets. When the angle exceeds 30°, the peak-to-peak voltage weakens. It is basically consistent with the change in magnetic field distribution. Between 0 and 60°, the peak-to-peak energy harvesting voltage and the magnetic field distribution are symmetrical at about the angle of 30°. It shows that when the angle between the iron sheet and the Halbach array rotor is 30°, the energy harvesting effect is the best.

3.2. The Influence of the Radian Size of the Iron Sheet and the Halbach Array

To explore the influence of the radian of the iron sheet on the Halbach array, the radian of the iron sheet is defined, as shown in Figure 6a.

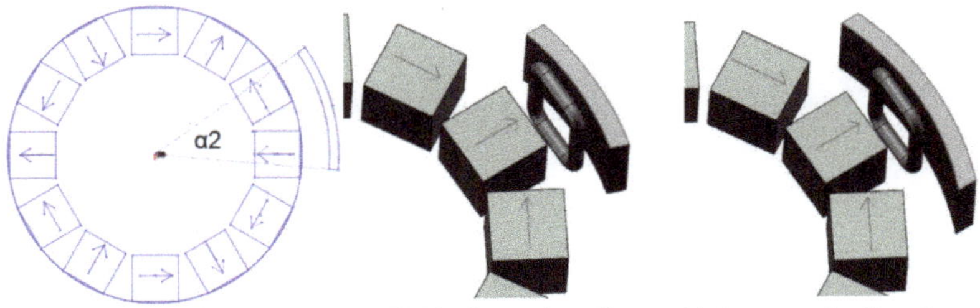

(**a**) Radian representation (**b**) No change in radian (**c**) Contrast of radian change

Figure 6. The radian of iron sheet.

It can be seen from the Figure 7b,c that the degree to which the iron sheet surrounds the Halbach array rotor is different, and it has been known from the above that when the angle between the iron sheet and the Halbach array is 30°, the energy harvesting effect is the best. Therefore, to explore the influence of the radian size of iron sheet on the Halbach array, the distance between the iron sheet and the Halbach array is 3 mm. The angle between the iron sheet and the Halbach array is 30°. The distribution of the Halbach array magnetic field under different radian of iron sheet is shown in Figure 7.

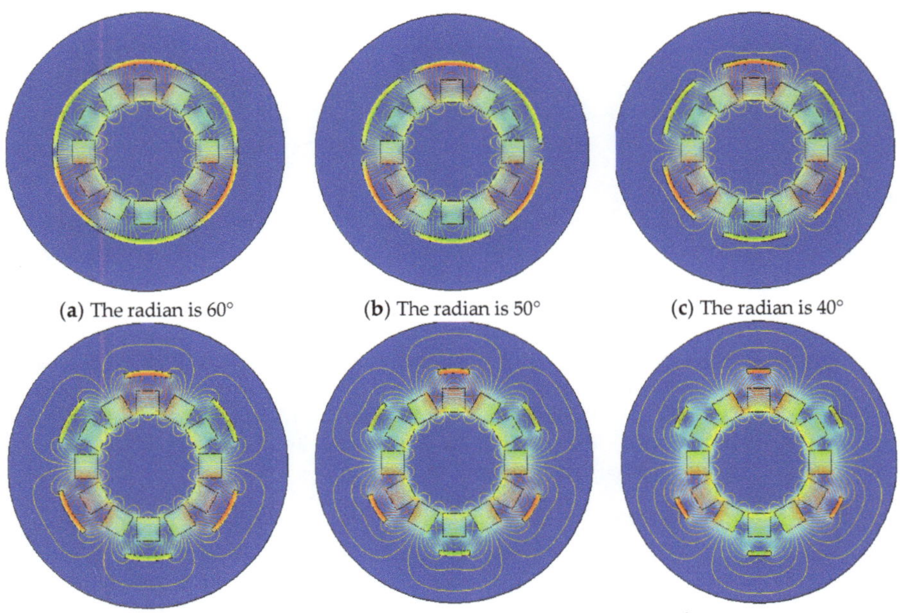

(**a**) The radian is 60° (**b**) The radian is 50° (**c**) The radian is 40°

(**d**) The radian is 30° (**e**) The radian is 20° (**f**) The radian is 15°

Figure 7. Magnetic field distribution of Halbach array under different iron sheet radians.

As you can see from Figure 7 above, when the angle between the Halbach array and the iron sheet is 30° and distance between the Halbach array and the iron sheet is 3 mm, the different sizes of iron sheet will result in different magnetic field distributions of the Halbach array. When the radian of each iron sheet is 60°, the iron sheet completely surrounds the periphery of the Halbach array rotor, the magnetic lines are compressed between the iron sheet and the Halbach array rotor, and the magnetic flux density modulus is the largest. When the radian of the iron sheet is 50°, it is the same as when the iron sheet is 60°. When the radian of iron sheet continues to decrease to 40°, the magnetic field distribution begins to change, iron sheet cannot completely surround the magnetic lines, and a small part of the magnetic lines stretch to the periphery of the iron sheet. When the radian of the iron sheet continues to decrease, the trend of change begins to increase, magnetic lines outside the iron sheet began to increase, and magnetic field density inside the iron sheet began to decrease. When the radian of the iron sheet is reduced to 15°, compared with the Halbach array magnetic field when there is no iron sheet, the magnetization effect only occurs in the iron sheet, and the magnetic flux density modulus is significantly reduced. In order to verify these conclusions, the energy harvesting simulation is carried out. Under the same conditions and under different iron radians, the maximum energy harvesting voltage obtained is shown in Figure 8.

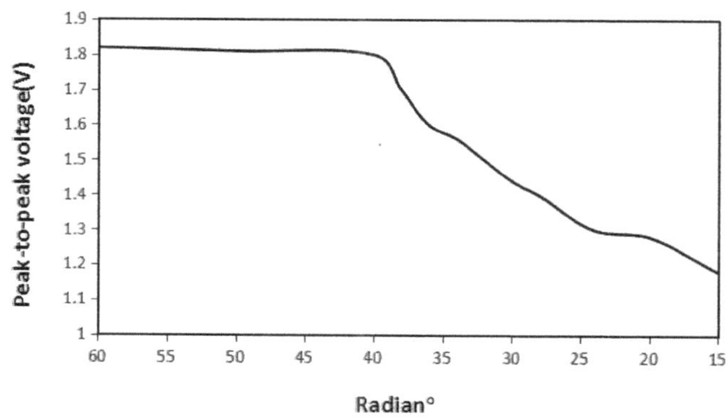

Figure 8. The maximum energy harvesting voltage of the Halbach array under different iron sheet radians.

It can be seen from Figure 8, when the radian of the iron sheet is between 60° and 40°, the energy harvesting effect does not change much, which is consistent with the magnetic field distribution results obtained above. The maximum energy harvesting voltage is 1.8 V. When radian of the iron sheet continues to decrease due to the weakening of the magnetic accumulation effect, its maximum energy harvesting voltage begins to decrease. The smaller the radian of the iron sheet, the lower its maximum energy harvesting voltage. When the radian of the iron sheet is reduced to 15°, its maximum energy harvesting voltage is 1.21 V. It shows that when the iron sheet and the Halbach array are at the same angle and distance, and radian of the iron sheet is 40°, the amount of iron sheet is the least and the energy harvesting effect is the best.

3.3. *The Influence of the Distance between the Iron Sheet and the Halbach Array*

The difference in the distance between the iron sheet and the Halbach array will lead to the difference in its magnetization effect. Distance between the iron sheet and the Halbach array is defined as the distance from the periphery of the Halbach array rotor to the iron sheet, as shown in Figure 9a.

It can be seen from the Figure 10b,c that only the distance between the iron sheet and the Halbach array is changed. When the angle between the fixed iron sheet and the Halbach array is 30°, and the radian of each iron sheet is 40°. If only changing the distance between the iron sheet and the Halbach array rotor, the magnetic field distribution diagram under different conditions is shown in Figure 10.

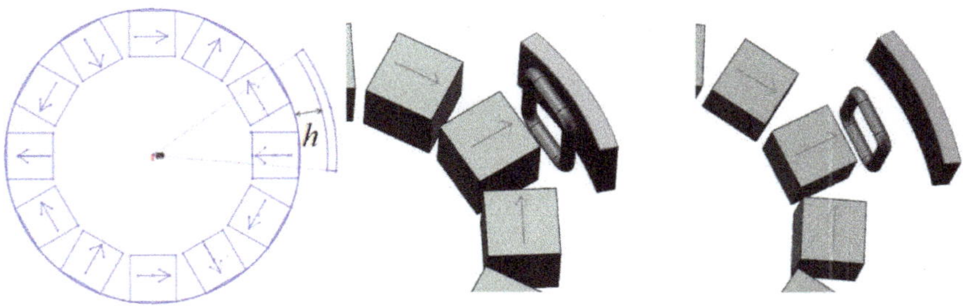

(a) Distance representation (b) No change in distance (c) Contrast of distance change

Figure 9. The distance between iron sheet and the Halbach array.

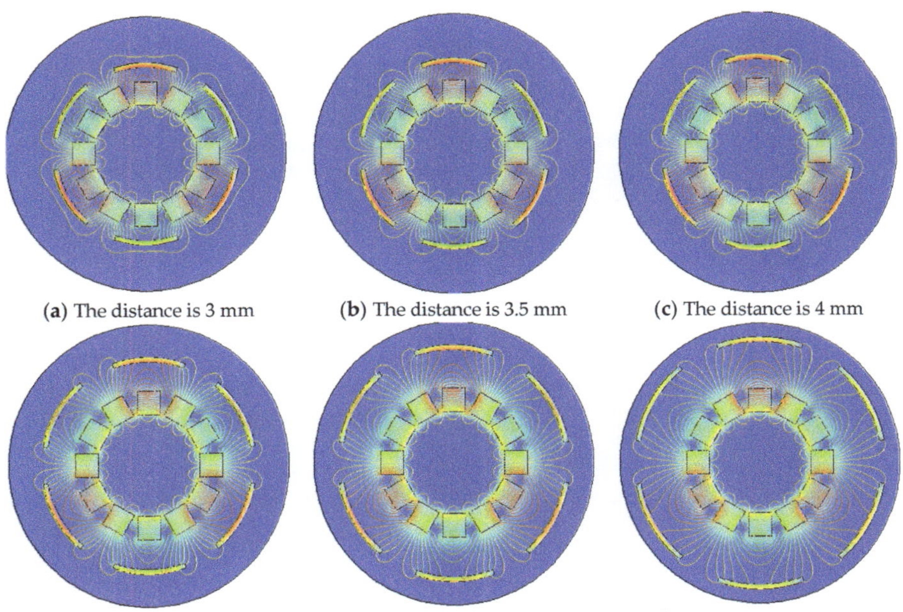

(a) The distance is 3 mm (b) The distance is 3.5 mm (c) The distance is 4 mm

(d) The distance is 6 mm (e) The distance is 8 mm (f) The distance is 10 mm

Figure 10. Magnetic field distribution of iron sheet and Halbach array at different distances.

It can be seen from Figure 10 that when the distance is 3 mm, the magnetic lines between the iron sheet and the Halbach array are highly concentrated and densely distributed. As the distance increases, the magnetic lines still gather between the iron sheet and the Halbach array. Still, the concentration of magnetic lines gradually decreases, and the distribution becomes more and more sparse. Compared with the distance of 3 mm, when the space is 10 mm, there is a significant difference in the density of magnetic lines. It shows that the distance between the iron sheet and the Halbach array does cause a difference in the magnetic field distribution. In order to verify the influence of the distance between the iron sheet and the Halbach array on the energy harvesting effect, the energy harvesting simulation was carried out under the same condition, and the maximum energy harvesting voltage at a different distance is shown in Figure 11.

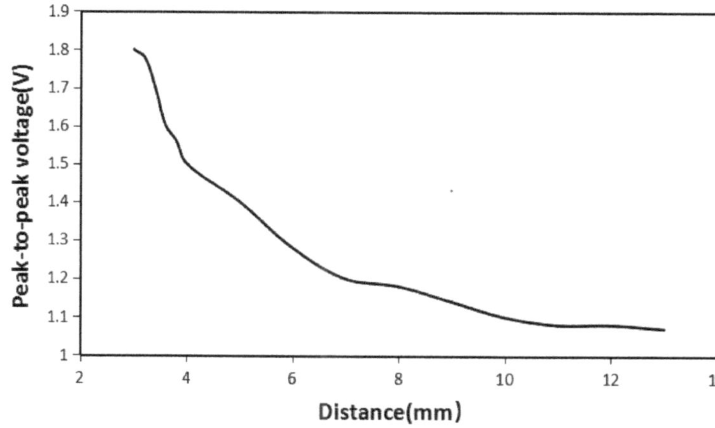

Figure 11. The maximum energy harvesting voltage of iron sheet and the Halbach array at different distance.

It can be seen from Figure 11 that as the distance between the iron sheet and Halbach array increases, the maximum energy harvesting voltage decreases. When the distance is 3 mm, the maximum energy harvesting voltage is 1.8 V, and the change in the distance and the maximum energy harvesting voltage is inversely proportional. When the distance just starts to increase, the maximum energy harvesting voltage drops extremely fast. When the distance continues to increase, the maximum energy harvesting voltage decreases slowly. When the distance is increased to 11 mm, the maximum energy harvesting voltage is basically maintained at about 1.07 V. Compared with the case without an iron sheet, the increase in energy harvesting effect is not obvious. It shows that in order to increase the energy harvesting effect, the distance between the iron sheet and the Halbach array should be reduced as much as possible.

4. Experimental Verification

From the above analysis, it can be known that the iron sheet has an important influence on the distribution of the magnetic field and energy harvesting effect of the Halbach array harvester. The angle between the iron sheet and the Halbach array, the radian of the iron sheet and the distance between the iron sheet and the Halbach array are all important for the energy harvesting effect. A set of comparative experiments was carried out. The coil used six groups of 100 turns each; the gap between the coil and the outside of the Halbach array rotor was 1 mm. The coil was located between the iron sheet and the Halbach array, and the distance between the iron sheet and Halbach array was 7 mm, and the rotation rate of Halbach array was 10 r/s. The experimental set up used in this paper is shown in Figure 12.

In order to verify the availability of energy harvesting of the Halbach array harvester, the Halbach array without iron sheets was selected to be tested through supplying energy to several low power devices, such as timers and thermometers. The arrangement of Halbach array and coils are shown in Figure 12 above. The gap between the coil and the outside of Halbach array rotor was 1 mm.

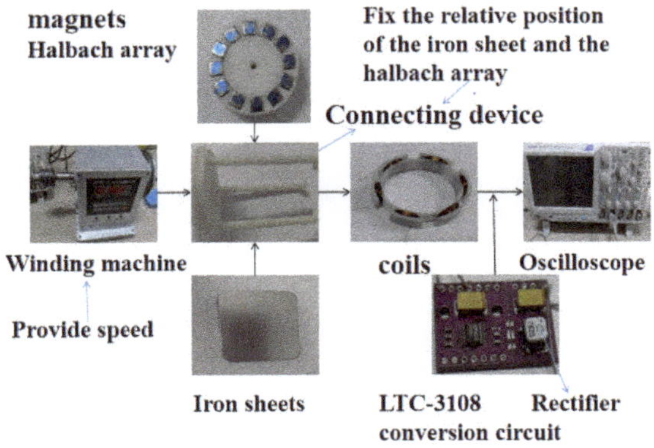

Figure 12. Devices used in the experiment.

As shown in Figure 13, the Halbach array harvester can fully meet the power supply for various low power devices.

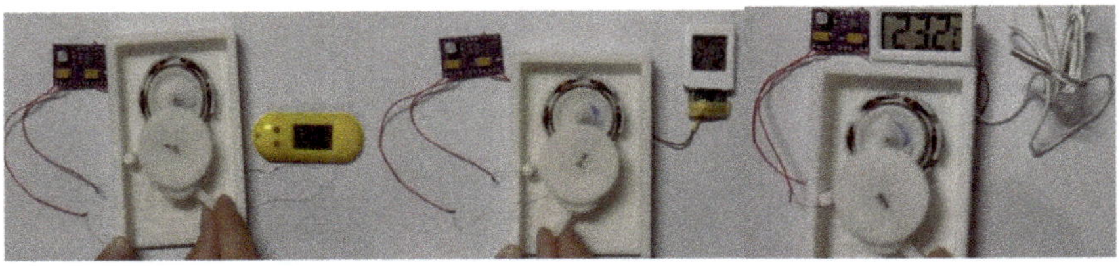

Figure 13. Energy supply experiment of Halbach array to different devices.

As can be seen from Figure 14, the measured voltage reached the maximum in extreme time without an iron sheet in the structure and when the angle between the fixed iron sheet and the Halbach array was 30° and the distance between the iron sheet and the Halbach array was 7 mm. The measured result verified the simulation results.

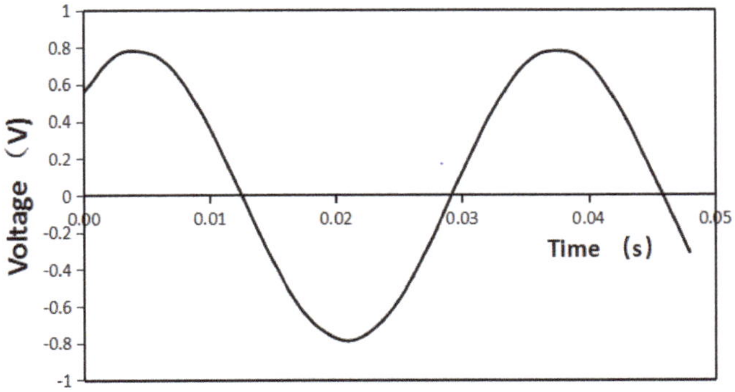

Figure 14. Voltage measured without iron sheet.

As can be seen from Figure 15, the relationship between the radian of the iron sheet and the energy harvesting voltage of the Halbach array is consistent with the simulation results. When the radian of the iron sheet was changed from 60 to 40°, the maximum energy harvesting voltage was basically unchanged. When the radian of the iron sheet continued to decrease, the maximum energy harvesting voltage also decreased. Through the transformation of the LTC-3108 circuit, the voltage can be stabilized to 2.8 V.

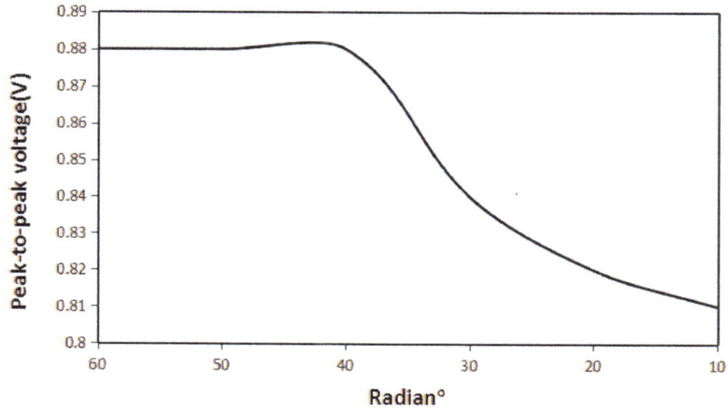

Figure 15. The relationship between the radian of iron sheet and the maximum voltage.

Figure 16 shows the time required to boost the voltage to 2.8 V under different radians of iron sheets. As can be seen from the figure, as the radian of iron sheet gets bigger, the time it takes to reach a stable voltage decreases. Therefore, iron sheet plays an important role in practical harvesting application.

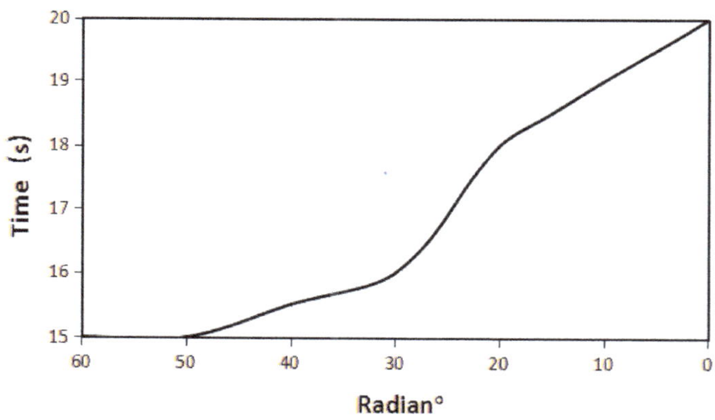

Figure 16. The time required to boost the voltage to 2.8 V under different iron radians.

5. Conclusions

Iron is an excellent ferromagnetic material. In this paper, simulations and experimental tests have proved that iron sheet indeed improves the energy harvesting effect of the Halbach array energy-harvesting structure. The energy-harvesting efficiency of the Halbach array can be significantly improved when iron sheet is added into the Halbach array at an appropriate angle and distance. When the angle between the iron sheet and the Halbach array is in the middle of the two radial magnetic poles, the Halbach array harvester with the iron sheet has the best energy-harvesting effect. The radian of the iron sheet determines the closure of the Halbach array. The larger the radian is, the higher the density magnetic lines are, and the better the energy-harvesting effect is. From completely enclosed to smaller and smaller radians, we found that the minimum radian of the iron sheet makes the least amount of sheet metal when the energy-harvesting effect is the greatest. The distance between the iron sheet and the Halbach array should be reduced as much as possible. With the decrease in distance, the energy-harvesting effect will increase. After adding the iron sheet the magnetic focusing effect of the Halbach array is enhanced, and the energy-harvesting efficiency is also improved and the structure volume is reduced. Because of its simple structure and easy miniaturization, it can be well applied to energy harvesting.

Author Contributions: Conceptualization, H.L. and X.Z.; methodology, T.P.; software, X.Z.; validation, X.Z., T.P. and Y.H.; investigation, Y.H.; data curation, X.Z.; writing—original draft preparation, X.Z.; writing—review and editing, B.S., H.G. All authors have read and agreed to the published version of the manuscript.

Funding: This research received no external funding.

Conflicts of Interest: The authors declare no conflict of interest.

References

1. Ren, Z.; Xu, J.; Le, X.; Lee, C. Heterogeneous wafer bonding technology and thin-film transfer technology-enabling platform for the next generation applications beyond 5G. *Micromachines* **2021**, *12*, 946. [CrossRef] [PubMed]
2. Lu, X.; Xi, X.; Lu, K.; Wang, C.; Chen, X.; Wu, Y.; Wu, X.; Xiao, D. Miniature ultralight deformable squama mechanics and skin based on piezoelectric actuation. *Micromachines* **2021**, *12*, 696. [CrossRef]
3. He, R.; Li, J. Optimized design and performance analysis of a magnetic-field modulated brushless dual-mechanical port motor with Halbach array permanent magnets. *Adv. Mech. Eng.* **2020**, *12*, 168781402095775. [CrossRef]
4. Sim, M.S.; Ro, J.S. Semi-analytical modeling and analysis of Halbach array. *Energies* **2020**, *13*, 1252. [CrossRef]
5. Zhang, H.S.; Deng, Z.X.; Yang, M.L.; Zhang, Y.; Tuo, J.Y.; Xu, J. Analytical prediction of Halbach array permanent magnet machines considering finite tooth permeability. *IEEE Trans. Magn.* **2020**, *56*, 1–10. [CrossRef]
6. Ni, Y.; Jiang, X.; Xiao, B.; Wang, Q. Analytical modeling and optimization of dual-layer segmented halbach permanent magnet machines. *IEEE Trans. Magn.* **2020**, *56*, 1–10. [CrossRef]
7. Xu, L.; Zhu, X.; Zhang, C.; Zhang, L.; Quan, L. Power oriented design and optimization of dual stator linear-rotary generator with halbach pm array for ocean energy conversion. *IEEE Trans. Energy Convers.* **2021**, *36*, 3414–3426. [CrossRef]
8. Jing, L.; Gong, J.; Chen, J.; Huang, Z.; Qu, R. A novel coaxial magnetic gear with unequal Halbach arrays and non-uniform air gap. *IEEE Trans. Appl. Supercond.* **2020**, *30*, 1–5. [CrossRef]
9. Chunsheng, E.; Smith, D.; Wolfe, J.; Weller, D.; Khizroev, S.; Litvinov, D. Physics of patterned magnetic medium recording: Design considerations. *J. Appl. Phys.* **2005**, *98*, 978–979.
10. Kou, B.; Zhang, L.; Zhao, B.; Zhang, C. Analysis and Optimization of a Linear Synchronous Motor with Novel Halbach Magnet Array. *Appl. Mech. Mater.* **2013**, *2698*, 416–417. [CrossRef]
11. Zhu, Z.Q.; Howe, D.; Chan, C.C. Improved analytical model for predicting the magnetic field distribution in brushless permanent-magnet machines. *IEEE Trans. Magn.* **2002**, *38*, 229–238. [CrossRef]
12. Sun, S. Monodisperse FePt nanoparticles and ferromagnetic FePt nanocrystal superlattices. *Cheminform* **2000**, *287*, 1989–1992. [CrossRef]
13. Zarko, D.; Ban, D.; Lipo, T.A. Analytical calculation of magnetic field distribution in the slotted air gap of a surface permanent-magnet motor using complex relative air-gap permeance. *IEEE Trans. Magn.* **2006**, *42*, 1828–1837. [CrossRef]
14. Gao, L.; Zhuang, J.; Nie, L.; Zhang, J.; Zhang, Y.; Gu, N.; Wang, T.; Feng, J.; Yang, D.; Perrett, S.; et al. Intrinsic peroxidase-like activity of ferromagnetic nanoparticles. *Nat. Nanotechnol.* **2007**, *2*, 577–583. [CrossRef]
15. Schmidt, G.; Ferrand, D.; Molenkamp, L.W.; Filip, A.T.; Van Wees, B.J. Fundamental obstacle for electrical spin injection from a ferromagnetic metal into a diffusive semiconductor. *Phys. Rev. B* **2000**, *62*, R4790. [CrossRef]
16. Zhu, Z.Q.; Xia, Z.P.; Atallah, K.; Jewell, G.W.; Howe, D. Powder alignment system for anisotropic bonded NdFeB Halbach cylinders. *IEEE Trans. Magn.* **2000**, *36*, 3349–3352. [CrossRef]
17. Xia, C.; Guo, L.; Wang, H. Modeling and analyzing of magnetic field of segmented Halbach array permanent magnet machine considering gap between segment. *IEEE Trans. Magn.* **2014**, *50*, 1–9. [CrossRef]
18. Tewari, S.; O'Reilly, T.; Webb, A. Improving the field homogeneity of fixed- and variable-diameter discrete Halbach magnet arrays for MRI via optimization of the angular magnetization distribution. *J. Magn. Reson.* **2021**, *324*, 106923. [CrossRef] [PubMed]
19. Žežulka, V.; Straka, P. Linear Halbach Structures:The Influence of Different Arrangement and Dimensions on the Resulting Magnetic Field. *J. Magn.* **2018**, *23*, 229–237. [CrossRef]
20. Lesnak, M.; Ciprian, D.; Les, M.; Lu, J. The effect of magnet magnetization vector orientation in Halbach arrays on the effectiveness of separation. *Acta Phys. Pol. A* **2014**, *126*, 158–159.
21. Liu, G.D.; Zhou, Y.F.; Zhou, R.G.; Huo, L.G. Modeling of the 2D Halbach array with different dimensions of magnets based on magnetization direction. *Appl. Mech. Mater.* **2015**, *752*, 396–401. [CrossRef]

micromachines

MDPI

Article

Time-Domain Dynamic Characteristics Analysis and Experimental Research of Tri-Stable Piezoelectric Energy Harvester

Xuhui Zhang [1,2,*], Luyang Chen [1], Xiaoyu Chen [1,3], Fulin Zhu [1] and Yan Guo [1]

1 College of Mechanical Engineering, Xi'an University of Science and Technology, Xi'an 710054, China; chenluyang@stu.xust.edu.cn (L.C.); 19105016004@stu.xust.edu.cn (X.C.); 20205224060@stu.xust.edu.cn (F.Z.); g1014901143@163.com (Y.G.)
2 Shaanxi Key Laboratory of Mine Electromechanical Equipment Intelligent Monitoring, Xi'an University of Science and Technology, Xi'an 710054, China
3 College of Engineering, Zunyi Normal College, Zunyi 563006, China
* Correspondence: zhangxh@xust.edu.cn

Abstract: In order to explore the dynamic characteristics of the linear-arch beam tri-stable piezoelectric energy harvester (TPEH), a magnetic force model was established by the magnetic dipole method, and the linear-arch composite beam nonlinear restoring force model was obtained through experiments. Based on the Euler–Bernoulli beam theory, a system dynamic model is established, and the influence of the horizontal distance, vertical distance and excitation acceleration of magnets on the dynamic characteristics of the system is simulated and analyzed. Moreover, the correctness of the theoretical results is verified by experiments. The results show that the system can be mono-stable, bi-stable and tri-stable by adjusting the horizontal or vertical spacing of the magnets under proper excitation. The potential well of the system in the tri-stable state is shallow, and it is easier to achieve a large-amplitude response. Increasing the excitation level is beneficial for the large-amplitude response of the system. This study provides theoretical guidance for the design of linear-arch beam TPEH.

Keywords: linear-arch beam; tri-stable piezoelectric energy harvester; nonlinear magnetic force; dynamic modeling

Citation: Zhang, X.; Chen, L.; Chen, X.; Zhu, F.; Guo, Y. Time-Domain Dynamic Characteristics Analysis and Experimental Research of Tri-Stable Piezoelectric Energy Harvester. *Micromachines* **2021**, *12*, 1045. https://doi.org/10.3390/mi12091045

Academic Editors: Kai Tao and Yunjia Li

Received: 24 July 2021
Accepted: 25 August 2021
Published: 29 August 2021

1. Introduction

In recent years, the application of wireless sensing technology in equipment monitoring, environmental monitoring and other fields has attracted more and more researchers' interest [1–3]. However, the battery life of wireless sensor nodes restricts its development, and the use of chemical batteries brings about high cost, environmental pollution and limited life span [4]. Obtaining energy from the environment has the potential to solve this problem and to address the power supply issue. Vibration energy has great advantages due to its ubiquity and high energy density. The piezoelectric vibration energy harvester has the advantages of a simple structure, compact size and easy integration and has broad application prospects [5].

In order to realize the collection and conversion of vibration energy, the piezoelectric cantilever beam has a simple structure and is prone to large strain after being excited. Therefore, domestic and foreign researchers have carried out a lot of research work on the cantilever beam piezoelectric energy harvester [6–9]. The traditional linear piezoelectric energy harvester can only work in a limited bandwidth where the environmental excitation is very close to its resonance frequency. When the environmental vibration frequency is far away from the resonance frequency of the device, the energy that can be captured is significantly reduced. This makes the energy harvester inefficient in practical applications, because most environmental vibrations occur randomly in a wide frequency range. Therefore, it is urgent to find a way to broaden the effective working frequency bandwidth to improve the performance of the vibration energy harvester [10]. The method of widening the working frequency bandwidth can be divided into two methods: linear frequency extension and nonlinear frequency extension [11]. The linear methods are array method [12], L-shaped beams [13], multi-degree-of-freedom beams [14] and nested structures [15]. Although these methods can effectively broaden the frequency band, the operating band width is still narrow for a single cantilever beam in these structures, and the system structure size is large. The efficiency of energy capture per unit volume is not high. Compared with the linear frequency extension method, the nonlinear technology can broaden the working frequency band of a single cantilever beam, thereby improving the power generation efficiency. Rezaei et al. [16] installed a spring on the free end of a linear cantilever beam. The research results show that the introduction of a nonlinear spring effectively broadens the resonance frequency band of the energy harvester. Liu et al. [17] designed an energy harvester with mechanical stoppers. The experimental results show that when the limiters are installed on both sides of the cantilever beam, the working frequency band can be broadened to 18 Hz.

In various nonlinear frequency extension structures and methods, the introduction of nonlinear magnetic force enables the energy harvester to work under nonlinear conditions of bi-stable, tri-stable and even more stable states [18–21]. Yao [22] proposed an L-beam bi-stable piezoelectric energy harvester. The research results show that the introduction of the end mass and nonlinear magnetic force increases the voltage and expands the effective frequency bandwidth. Wang [23] conducted research on the nonlinear magnetic force modeling of the tri-stable piezoelectric energy harvester, proposed an improved magnetic force model to describe the nonlinear magnetic force, and explored the influence of the magnet spacing on the magnetic force. Zhu [24] and Zhou [25] compared and analyzed the performance of the tri-stable piezoelectric energy harvester and the bi-stable piezoelectric energy harvester from the perspective of the frequency domain response characteristics of the energy harvester. The research results show that the tri-stable piezoelectric energy harvester has a shallower potential well, a wider trapping energy band and a higher output. Zhang [26] et al. designed a combined beam type tri-stable piezoelectric energy harvester, established a distributed parameter model based on the Hamiltonian principle, and analyzed the influence of the horizontal distance of the magnet and the excitation acceleration on the amplitude-frequency response characteristics from the perspective of the frequency domain. In order to improve the energy trapping efficiency, Li [27] proposed a tri-stable piezoelectric energy harvester with a trapezoidal well. Through the asymmetrical arrangement of the external magnet of the cantilever beam, the research results show that this design makes the large response easier to be triggered. Wang [28] derives the distributed parameter model of the energy harvester on the basis of considering the geometric nonlinearity of the cantilever beam (GNL) and the gravitational effect (GE). Theoretical and experimental studies have shown that the tri-stable energy harvester considering GNL and GE has an asymmetric potential well, which can improve the output performance. Unlike the conventional tri-stable energy harvester structure, Sun [29] also proposed a new type of tri-stable piezoelectric energy harvester with only a ring magnet outside, which can present different dynamic characteristics by adjusting the distance of the magnet. Most of the abovementioned energy harvesters that introduce a nonlinear magnetic field use classic straight beam, which cannot effectively collect multidirectional vibration energy in the environment, which limits the application of energy harvester in actual environments.

Aiming at the power supply problem of wireless monitoring nodes in coal mines and considering the characteristics of multidirectional and low-frequency excavation excitation, Zhang [30] proposed a new type of multifield coupled multidirectional piezoelectric energy harvester. As shown in Figure 1, the device can adapt to the mining environment and effectively collect vibration energy in different directions. Based on this, this paper designs a linear-arch beam TPEH. Under u-direction excitation conditions, a linear-arch beam

TPEH dynamic model is established, and the influence of the horizontal distance, vertical distance and excitation acceleration of the magnet on the dynamic response characteristics is analyzed by means of numerical simulation. An experimental platform was built to verify the correctness of the theoretical analysis, the research provided theoretical guidance for the design of the linear-arch beam TPEH. The structure of this article is arranged as follows: Section 2 provides a schematic diagram of the structure of the TPEH in this study and then introduces the establishment of the nonlinear restoring force model and the magnetic force model, and the dynamic model is obtained. In Section 3, the system potential energy under different magnet distances and the dynamic performance under different magnetic distances and excitation accelerations are analyzed through simulation. In Section 4, experiments were performed to verify the theoretical results. Finally, in Section 5, the main findings and conclusions are presented and summarized.

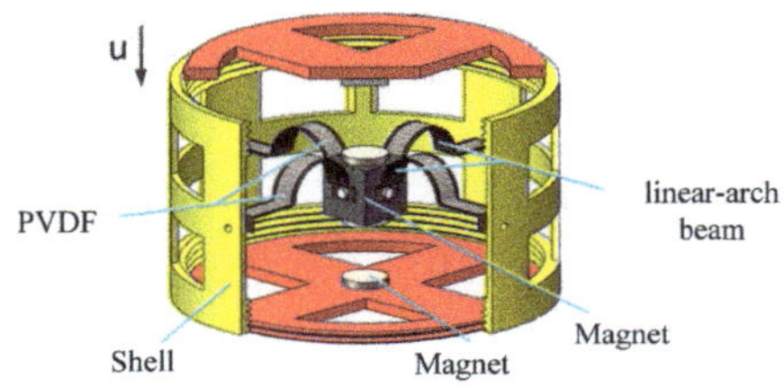

Figure 1. Multifield coupled multidirectional piezoelectric energy harvester.

2. Structure and Theoretical Model of a TPEH

2.1. Structure of the TPEH

Figure 2 shows the principle diagram of the structure of the TPEH. The system consists of a linear-arch composite beam, piezoelectric materials and three magnets. The linear-arch composite beam is pasted with flexible piezoelectric material PVDF; magnet A is fixed at the end of the cantilever beam; magnets B and C are symmetrically arranged on both sides of the x-axis; the horizontal distance between magnet A and B or C is d; and the vertical distance between magnets B and C is $2d_g$. In the figure, the length of the composite beam in the x-axis direction is L; the width of the composite beam is w; the thickness is h_s; the length of the linear beam is L_1; and the radius and chord length of the arched part are r and $2r$.The width of the PVDF pasted on the composite beam is the same as that of the composite beam; the length is L_2; and the thickness is h_p.

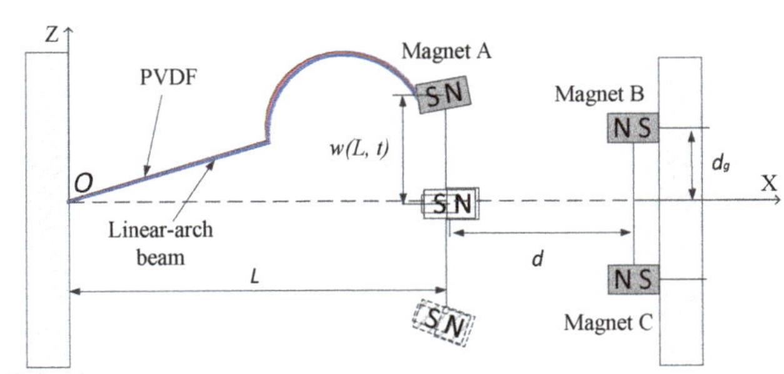

Figure 2. Schematic diagram of linear-arch beam TPEH.

2.2. Theoretical Modeling

2.2.1. Nonlinear Restoring Force of Linear-Arch Beam

Unlike the linear restoring force of the typical straight beam, the restoring force is nonlinear due to the arched structure in the linear-arched beam. The YLK-10 dynamometer is used to measure the restoring force of the composite beam structure in the z-axis direction, the average of multiple measurements is taken, and the curve-fitting method is used to obtain multiple expressions of the nonlinear restoring force

$$F_r = s_3 w(L,t)^3 + s_2 w(L,t)^2 + s_1 w(L,t) \quad (1)$$

where s_1, s_2 and s_3 are polynomial coefficients, and $w(L,t)$ is the displacement of the beam along the z-axis at time t.

Figure 3 shows the experimental measurement and curve-fitting results of the linear-arch beam's nonlinear restoring force. After fitting, the polynomial coefficients $s_1 = -14$ N/m, $s_2 = 254.586$ N/m^2, $s_3 = -56,681.2$ N/m^3. It can be observed from the figure that the restoring force of the beam is a curve due to the existence of the arched part. Taking $w(L,t) = 0$ as the equilibrium position of the cantilever beam, the restoring forces of the beam are asymmetric. This is because in the process of beam deformation, when the curvature of the arched part becomes larger, the force required is smaller than when the curvature becomes smaller. In the measurement results of the restoring force of the composite beam, the restoring force on both sides of the balance point is not completely symmetrical.

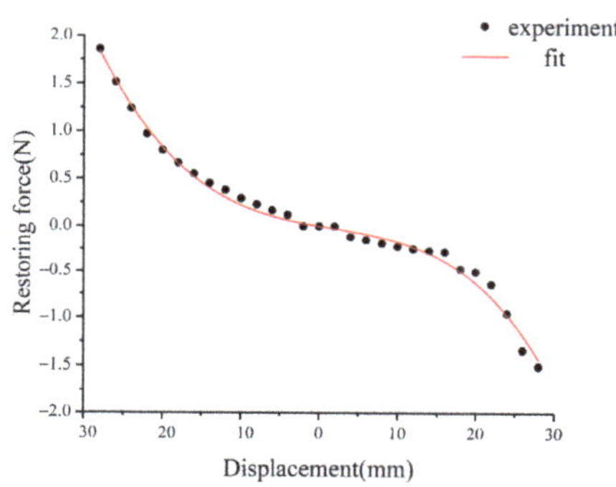

Figure 3. Nonlinear restoring force of linear-arch beam.

To determine the vibration displacement $w(x,t)$ of the beam, the Rayleigh-Ritz method is used to expand the vibration displacement of the composite beam as

$$w(x,t) = \sum_{i=1}^{n} \varphi_i(x) q_i(t) \quad (2)$$

where i is the order of the vibration mode of the composite beam, $\varphi_i(x)$ represents the i-th modal function of the composite beam, and $q_i(t)$ represents the i-th generalized modal coordinate. For the linear-arch beam in this paper, one end is clamped and fixed on the base, and the other end is free. The allowable function can be expressed as [31]

$$\varphi_i(x) = 1 - \cos\left[\frac{(2i-1)\pi x}{2L}\right] \quad (3)$$

Since the excitation of piezoelectric energy harvester is mainly low frequency, the first-order modal bending vibration of the beam plays a leading role; therefore, this paper only considers the first-order mode of the beam.

2.2.2. Nonlinear Magnetic Force Modeling

In order to accurately analyze the vibration characteristics of a piezoelectric cantilever beam, it is necessary to determine the magnitude of the nonlinear magnetic force at its end. The geometric relationship between magnets A, B and C is shown in Figure 4. This paper uses a magnetic dipole model to describe the nonlinear magnetic force.

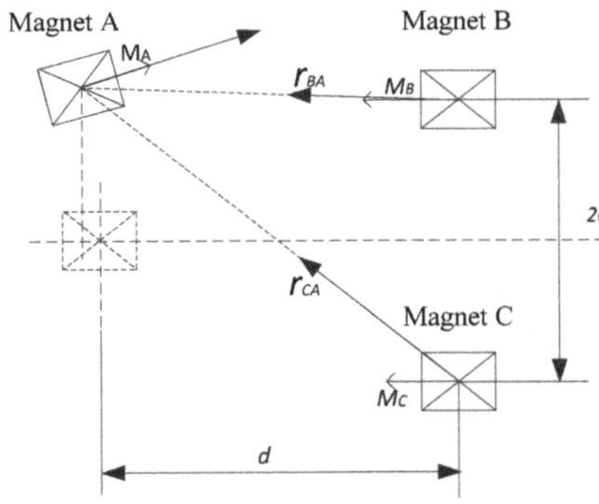

Figure 4. Nonlinear magnetic force model.

The magnetic flux density generated by magnet B at magnet A is

$$B_{BA} = -\frac{\mu_0}{4\pi}\nabla\frac{M_B\vec{r_{BA}}}{\|\vec{r_{BA}}\|_2^3} \tag{4}$$

where μ_0 is the vacuum permeability, ∇ is the vector gradient, $\vec{r_{BA}}$ is the direction vector from magnet B to A, and M_B is the magnetic moment of the magnetic dipole B. Then, the potential energy generated by the magnet B at the magnet A is

$$U_{BA} = -B_{BA}\cdot M_A \tag{5}$$

Compared with the composite beam size, the magnet size is smaller; therefore, $l_a sin\alpha \ll w(L,t)$, so $\Delta x \cong 0$ [32], so

$$\begin{aligned} r_{BA} &= -d\vec{i} + \left[w(L,t) - d_g\right]\vec{j} \\ M_A &= m_A V_A cos\alpha\,\vec{i} + m_A V_A sin\alpha\,\vec{j} \\ M_B &= -m_B V_B\,\vec{i} \end{aligned} \tag{6}$$

where $\vec{i}$ and $\vec{j}$ are the unit vectors in the x and z axis directions, respectively; m_A and m_B represent the magnetization of magnets A and B, respectively; and V_A and V_B represent the volumes of magnets A and B. Since $\alpha = \arctan[w'(L,t)]$, we have

$$cos\alpha = \frac{1}{\sqrt{[w'(L,t)]^2+1}}, sin\alpha = \frac{w'(L,t)}{\sqrt{[w'(L,t)]^2+1}} \tag{7}$$

Substituting Formulaes (4), (6) and (7) into Formula (5)

$$U_{BA} = \frac{\mu_0 m_A V_A m_B V_B * \{-\left[w(L,t)-d_g\right]^2 + 2d^2 - 3d\left[w(L,t)-d_g\right]w'(L,t)\}}{4\pi\sqrt{[w'(L,t)]^2+1} * \{d^2 + \left[w(L,t)-d_g\right]^2\}^{5/2}} \tag{8}$$

in the same way, the potential energy generated by magnet C at magnet A is as follows:

$$U_{CA} = \frac{\mu_0 m_A V_A m_C V_C * \{-\left[w(L,t)+d_g\right]^2 + 2d^2 - 3d\left[w(L,t)+d_g\right]w'(L,t)\}}{4\pi\sqrt{[w'(L,t)]^2+1} * \{d^2 + \left[w(L,t)+d_g\right]^2\}^{5/2}} \tag{9}$$

The potential energy generated by magnets B and C at magnet A at the end of the beam is

$$U_m = U_{BA} + U_{CA} \tag{10}$$

The magnetic force on magnet A is

$$F_m = \frac{\partial U_m}{\partial q(t)} \tag{11}$$

Using polynomial fitting to simplify the magnetic force to a polynomial about the displacement $w(L,t)$, we can obtain

$$F_M = K_1 w(L,t)^5 + K_2 w(L,t)^3 + K_3 w(L,t) \tag{12}$$

where K_1, K_2 and K_3 are obtained by curve fitting.

2.2.3. Dynamic Modeling

In this paper, Lagrange's equation is used to establish the motion equation of the linear-arch beam

$$L_a(x,t) = T_S + T_P + T_M + W_P - U_r - U_M \tag{13}$$

where T_S is the kinetic energy of the metal base layer, T_P is the kinetic energy of the piezoelectric layer, T_M is the kinetic energy of the end magnet, W_P is the electrical energy of the piezoelectric layer, and U_r is the potential energy of the linear-arch piezoelectric cantilever beam.

$$T_S = \frac{1}{2}\rho_S A_S \int_0^L \left(\dot{w}(x,t) + \dot{z}(t)\right)^2 dx \tag{14}$$

$$T_P = \frac{1}{2}\rho_P A_P \int_0^L \left(\dot{w}(x,t) + \dot{z}(t)\right)^2 dx \tag{15}$$

$$T_M = \frac{1}{2}M_t\left(\left[\dot{w}(x,t)\right]_{x=l} + \dot{z}(t)\right)^2 + \frac{1}{2}I_t\left(\left[\frac{\partial^2 w(x,t)}{\partial t \partial x}\right]_{x=l}\right)^2 \tag{16}$$

$$W_P = \frac{1}{2}\int_{V_{0P}} E_3 D_3 dV_P = \frac{1}{4}e_{31} b(h_S + h_P) v(t) \left[\frac{\partial w(x,t)}{\partial x}\right]_{x=l} + \frac{1}{2}C_P v^2(t) \tag{17}$$

$$U_r = \int F_r \, dq(t) \tag{18}$$

where "." means derivation with respect to t and "′" means derivation with respect to x. A_s is the cross-sectional area of the metal base layer, A_P is the cross-sectional area of the piezoelectric layer, $w(x,t)$ is the vibration displacement of the composite beam, $z(t)$ is the vibration displacement of the base, M_t is the mass of the magnet, and I_t is the moment of inertia of the magnet.

According to the Euler–Bernoulli beam theory and Kirchhoff's law, the system dynamics equations of linear-arch composite beams can be obtained

$$M\ddot{q}(t) + C\dot{q}(t) + F_r - \theta v(t) + F_M = -H\ddot{Z}(t) \tag{19}$$

$$\theta \dot{q}(t) + C_P \dot{v}(t) + \frac{v(t)}{R} = 0 \tag{20}$$

where

$$M = (\rho_S A_S + \rho_P A_P)\int_0^L (\varphi(x))^2 dx + M_t(\varphi(L))^2 + I_t(\varphi'(L))^2 \tag{21}$$

$$\theta = \frac{1}{2}e_{31} b(h_S + h_P)\varphi'(L) \tag{22}$$

$$H = (\rho_S A_S + \rho_P A_P)\int_0^L \varphi(x) dx + M_t \varphi(L) \tag{23}$$

The following nondimensionalized parameters are introduced [33]:

$$x(\tau) = \frac{q(t)}{l},\ \mu(\tau) = \frac{v(t)}{e},\ \tau = \omega_0 t,\ e = \frac{L\theta}{C_p} \tag{24}$$

where l is the length coefficient; e is the voltage coefficient; τ is the time coefficient; and $\omega_0 = \sqrt{\frac{K}{M}}$.

Incorporating Equation (24) into Equations (19) and (20), the dimensionless dynamic equation of the system is

$$\ddot{x} + 2\zeta\dot{x} - \vartheta\mu + \delta(F_m + F_r) = -\beta sin(\Omega\tau) \tag{25}$$

$$\dot{x} + \dot{\mu} + \alpha\mu = 0 \tag{26}$$

where $\zeta = \frac{C}{2\omega_0 M}$, $\vartheta = \frac{\theta^2}{KC_p}$, $\alpha = \frac{1}{\omega_0 RC_p}$, $\delta = \frac{1}{KL}$, $\beta = \frac{HA}{KL}$, $\Omega = \frac{\omega}{\omega_0}$.

Let, $x_1 = x$, $x_2 = \dot{x}$, $x_3 = u$. We can obtain the state space equation:

$$\begin{cases} \dot{x_1} = x_2 \\ \dot{x_2} = -2\xi x_2 - \vartheta x_3 - \delta(F_m + F_r) - \beta sin(\Omega\tau) \\ \dot{x_3} = -x_2 - \alpha x_3 \end{cases} \tag{27}$$

3. Dynamic Analysis of TPEH

3.1. Analysis of System Potential Energy

Table 1 shows the structural parameters of the beam. The total potential energy P of the system can be expressed as

$$P = U_m - U_b \tag{28}$$

Table 1. Structure and material parameters of TPEH.

Parameter	Value	Parameter	Value
linear-arch beam		Piezoelectric layer	
$L_1 * w * h_s\,(\mathrm{mm}^3)$	$20 * 8 * 0.2$	$L_2 * w * h_s\,(\mathrm{mm}^3)$	$40 * 8 * 0.2$
r (mm)	10	Permittivity constant(F/m)	110×10^{-12}
Young's modulus $(\mathrm{N/m^2})$	128×10^9	Young's modulus $(\mathrm{N/m^2})$	3×10^9
Density $(\mathrm{kg/m^3})$	8300	Density $(\mathrm{kg/m^3})$	1780

According to Formula (28), the system parameters of horizontal magnetic distance d and vertical magnetic distance d_g play a decisive role in the potential energy of the system. This section uses simulation to analyze the influence of parameters on the potential energy of the system.

Figure 5a shows the simulation results of the potential energy of the TPEH when $d_g = 8$ mm and d is 7, 11, 13, 15 and 22 mm, respectively. It can be observed that with the decrease in d, the potential-energy curve changes from a single potential well to a triple potential well, and the depth of the potential well gradually increases, because the magnetic force gradually increases with the gradual decrease in d.

Figure 5b shows the potential energy simulation results of the TPEH, when $d = 16$ mm, and d_g is 0, 6, 8, 12 and 25 mm. With the increase in d_g, the system goes through three motion states of bi-stable, tri-stable and mono-stable in sequence. When $d_g = 0$ mm, the magnets B and C coincide, and the potential energy curve of the system has two potential wells. With the gradual increase in d_g, the potential energy curve of the system gradually changes from two potential wells to three potential wells. With the increase in d_g, the depth of the potential well in the middle of the system becomes deeper, and the width increases; while the depth of the potential wells on both sides becomes shallower, the width is reduced. As dg continues to increase to 25 mm, the force between magnets A, B, and C is very small and almost zero. The system has only a single potential well, which appears as a mono-stable system. It can be observed from Figure 4a,b that especially when there are three potential wells in the potential energy curve, the two external potential wells are asymmetrical: one is high, and the other is low. This is due to the asymmetric potential well caused by the asymmetry of the restoring force.

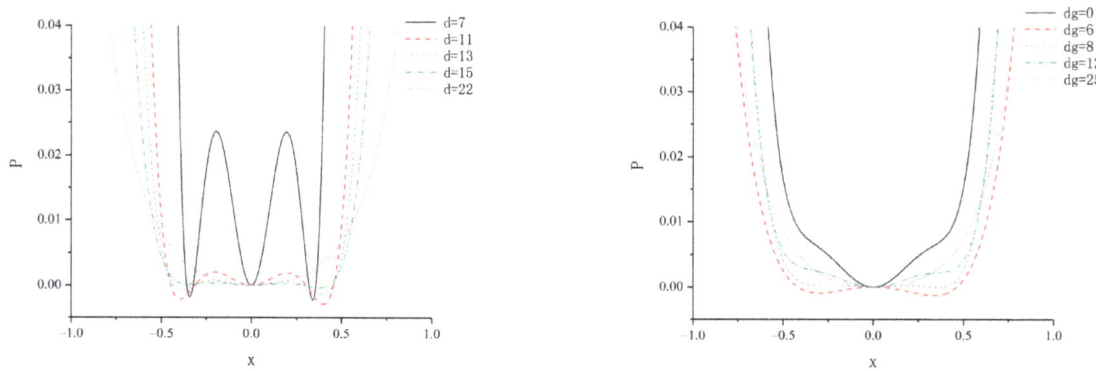

(a) (b)

Figure 5. System potential energy of different magnetic distance: (**a**) change the horizontal magnetic distance d; (**b**) change the vertical magnetic distance d_g.

3.2. Dynamic Analysis

Combining the results of the potential energy analysis, it can be observed that the magnet spacing has a significant effect on the system dynamics. This section explores the influence of the magnet horizontal spacing d, vertical spacing d_g and excitation acceleration a on the system dynamics characteristics.

3.2.1. The Influence of Horizontal Magnetic Distance d

When $A = 12\ \mathrm{m/s^2}$,$d_g = 8$ mm, $\Omega = 0.5$, changing the horizontal magnetic distance d, the displacement–velocity phase diagram and time-displacement history diagram of the TPEH are shown in Figure 6.

As shown in Figure 6a, when the horizontal distance between the magnets is $d = 22$ mm, the force between the magnets is small and has almost no effect on the beam. Therefore, the piezoelectric energy harvester exhibits mono-stable motion characteristics. With the decrease in d, the force between the magnets increases, and the influence of magnetic force on the motion characteristics of the energy harvester gradually appears, and the system changes from mono-stable to tri-stable. When $d = 15$ mm, the system can cross the potential barrier, the beam moves back and forth between the three equilibrium points. From Figure 6b, it can be observed that the system exhibits tri-stable motion characteristics, reaching a large-amplitude response; the dimensionless displacement is 0.45; and the response displacement reaches 18 mm. When the horizontal distance between the magnets is reduced to 13 mm, due to the large force between the magnets, the beam struggles to get rid of the restraint of the magnetic force. As shown in Figure 6c, the system exhibits mono-stable motion characteristics. The end of the beam makes a small periodical movement near the center balance point, and the system response displacement and output voltage are very small.

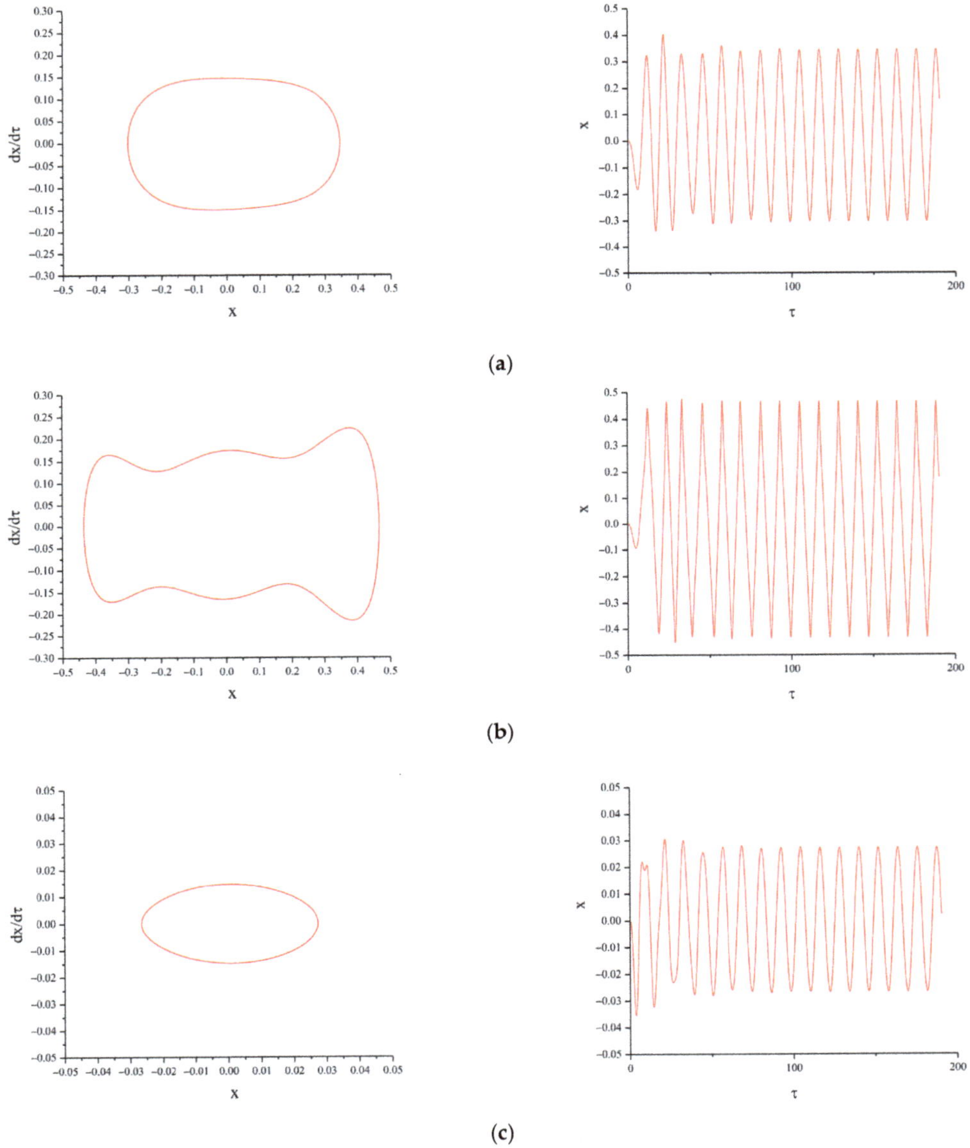

Figure 6. Phase portrait and time-displacement diagram of different magnetic distance d: (**a**) $d = 22$ mm; (**b**) $d = 15$ mm; (**c**) $d = 13$ mm.

3.2.2. The Influence of Vertical Magnetic Distance d_g

When $A = 12\ \mathrm{m/s^2}$, $d = 16\mathrm{mm}$, $\Omega = 0.5$, changing the magnet vertical distance d_g, the displacement-velocity phase diagram and time history diagram of the TPEH are shown in Figure 7.

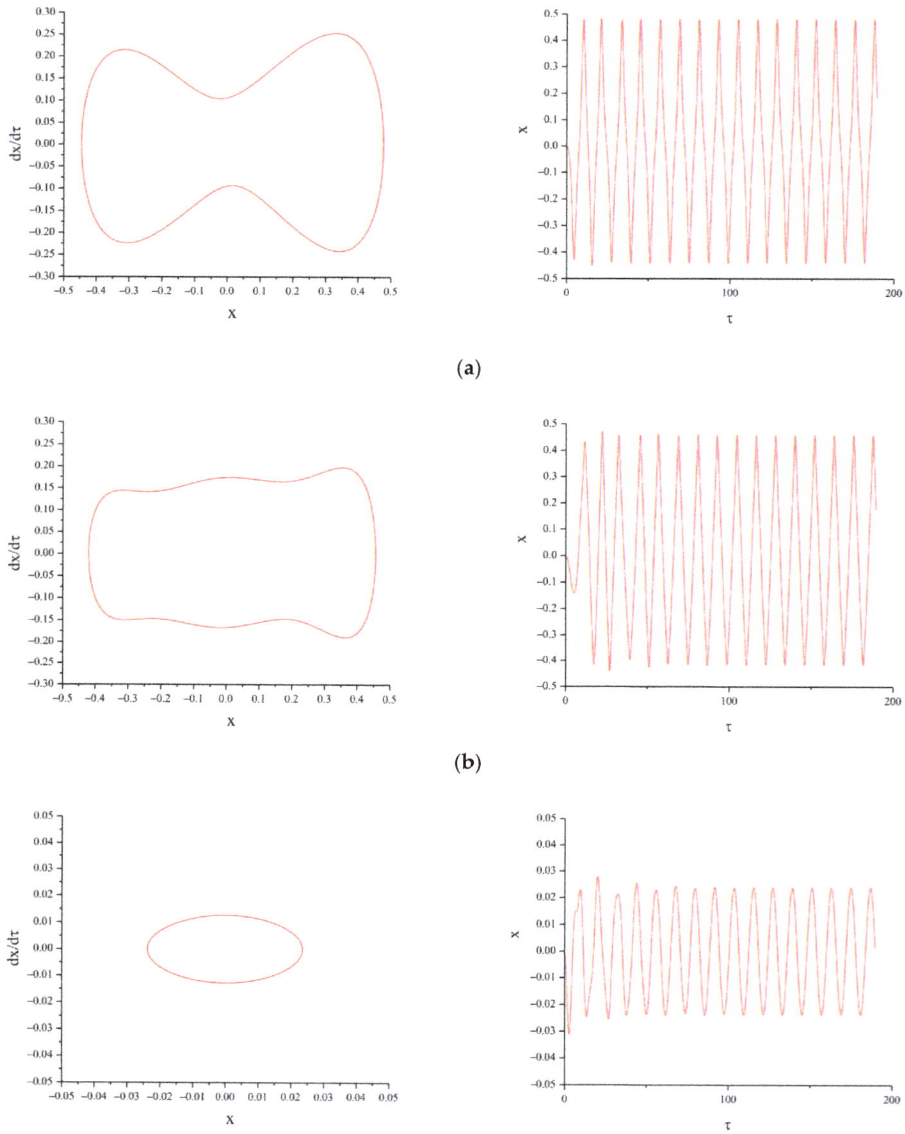

Figure 7. Phase portrait and time-displacement diagram of different magnetic distance d_g: (**a**) $d_g = 6$ mm; (**b**) $d_g = 8$ mm; (**c**) $d_g = 12$ mm.

When the vertical distance between the magnets B and C is small, the system potential energy curve has two potential wells. When the excitation level is low, the system cannot cross the potential barrier and can only move in the well; therefore, the system cannot exhibit bi-stable characteristics. From the Figure 7, it can be observed that as d_g increases, the system goes through bi-stable, tri-stable, and mono-stable in sequence. As shown in Figure 7a, when $d_g = 6$ mm, bi-stable motion can be achieved. As shown in Figure 7b, when $d_g = 8$ mm, the system can easily cross the potential barrier and move back and forth between the three steady-state positions, showing a tri-stable characteristic. At this time, the displacement and speed of the end of the beam are greater than other states, and the response displacement reaches 18 mm. As d_g continues to increase, the force between the magnets is small, the system potential energy has only one potential well, and there is no multistable characteristic. Under low-level excitation, the system response displacement and velocity are both small.

3.2.3. The Influence of Excitation Intensity A

When $d = 15$ mm, $d_g = 8$ mm, $\Omega = 0.5$, the excitation acceleration a is changed. Figure 8 shows the displacement-velocity phase diagrams with accelerations of 5 $\mathrm{m/s^2}$, 7 $\mathrm{m/s^2}$ and 12 $\mathrm{m/s^2}$.

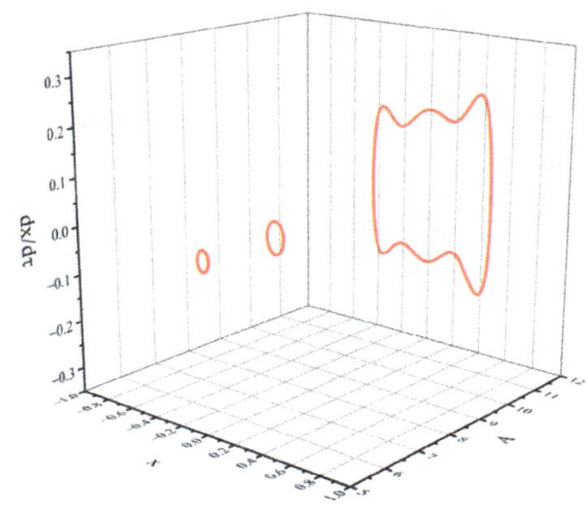

Figure 8. Phase portrait of different excitation acceleration A.

From Figure 8, we can see that: under the above conditions, the system potential energy curve has three potential wells. When the excitation is low, the system cannot cross the potential barriers on both sides and can only make periodic motions in the middle potential well. When $A = 5\ \mathrm{m/s^2}$, the system moves in a mono-stable state, with *a* small periodic vibration centered on the intermediate balance point. Increasing the excitation acceleration, the energy obtained by the system increases. When $A = 7\ \mathrm{m/s^2}$, the displacement of the beam increases, but the excitation magnitude is still not enough to make the system cross the two barriers. When the excitation acceleration increases to 12 $\mathrm{m/s^2}$, the system presents a tri-stable motion characteristic, moving periodically between the three potential wells. Combining the previous analysis of the influence of the variable magnetic distance on the system dynamics, when the system presents a bi-stable or tri-stable state of motion, the steady-state points on both sides of the displacement-velocity phase diagram are asymmetrical. The speed on one side is higher than the other side, which is caused by the asymmetry of the restoring force of the linear-arch beam.

4. Experimental Validation

In order to verify the correctness of the theoretical analysis of the linear-arch beam tri-stable piezoelectric energy harvester, an experimental platform was built for experimental verification. The experimental device is shown in Figure 9: the linear-arched beam is fixed on the base by a clamp and is perpendicular to the horizontal sliding table of the vibrating table. Two external magnets B and C are fixed on the base, mutually exclusive with the magnet A. In the experiment, the excitation signal is set by the computer, and the sinusoidal signal is sent out by the vibration controller (VT-9008), which is amplified by the power amplifier (GF-20) and output to the vibration table (E-JZK-5T). The vibration table operates according to the preset excitation signal. Through the reflective sticker on the top of the arched section beam, the laser vibrometer (LV-S01, resolution: 1 um/s) can measure the speed of the cantilever, by using a handheld vibrometer (coco80, sampling rate: 2 KHz) to collect data.

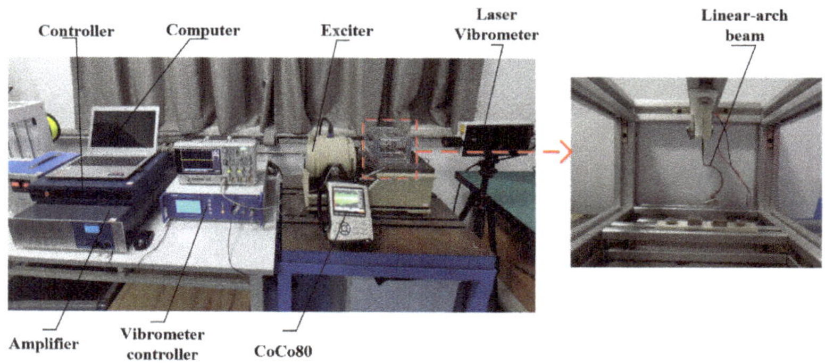

Figure 9. Experimental platform and TPEH.

Figure 10 is the experimental results of displacement-velocity of linear-arch composite beams with different excitation accelerations when $d = 15$ mm, $d_g = 8$ mm and $f = 8$ Hz. As shown in Figure 10a, when $A = 5\ \mathrm{m/s^2}$, the end displacement is 2 mm. As the excitation acceleration increases, the displacement amplitude increases. When $A = 7\ \mathrm{m/s^2}$, the end displacement increases to 2.5 mm. It shows that when the excitation is low, the system cannot cross the barriers on both sides and can only move in the well. Figure 10c is the displacement-velocity phase diagram at $12\ \mathrm{m/s^2}$. The system can cross the potential barrier and show the characteristics of tri-stable motion. As shown in Figure 11, the three stable positions of the tri-stable system can be clearly seen.

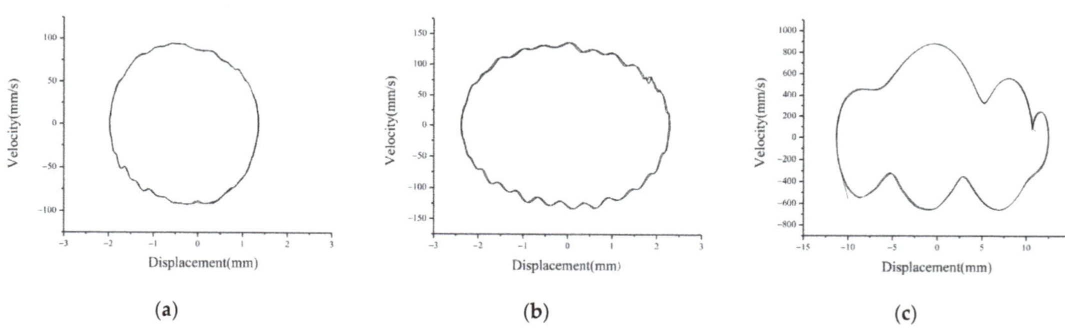

Figure 10. Experimental phase portrait of different excitation accelerations: (**a**) a = $5\ \mathrm{m/s^2}$; (**b**) a = $7\ \mathrm{m/s^2}$; (**c**) a = $12\ \mathrm{m/s^2}$.

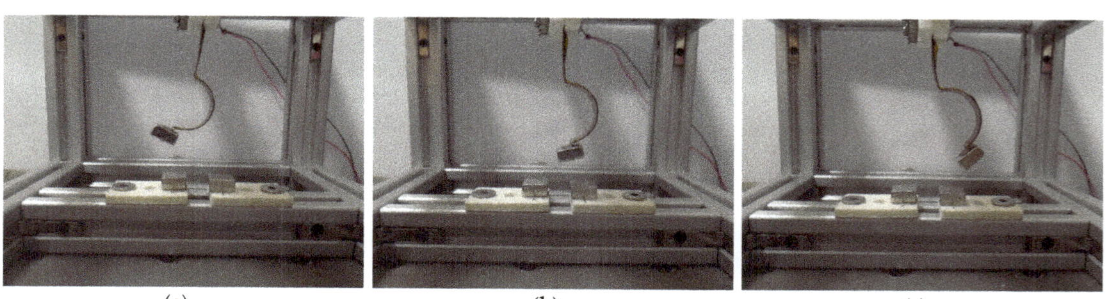

Figure 11. Three stable equilibrium positions: (**a**) stable1; (**b**) stable2; (**c**) stable3.

Figure 12 shows the experimental results of displacement-velocity of linear-arched beam under different magnetic distances when $f = 8$ Hz and $A = 12\ m/s^2$. As shown in Figure 12a, when $d = 22$ mm, $d_g = 8$ mm, the system has only one potential energy well, and the system moves periodically in the well. As shown in Figure 12b, when $d = 16$ mm, $d_g = 6$ mm, the system has two potential wells, and the excitation intensity is sufficient for enabling the system to cross the potential barrier to achieve bi-stable motion. Figure 13 shows the two stable positions of the system. When $d = 16$ mm, $d_g = 8$ mm, as shown in Figure 12c, the system cyclically moves between three stable positions, showing three-stable characteristics, and the displacement amplitude is greatly increased.

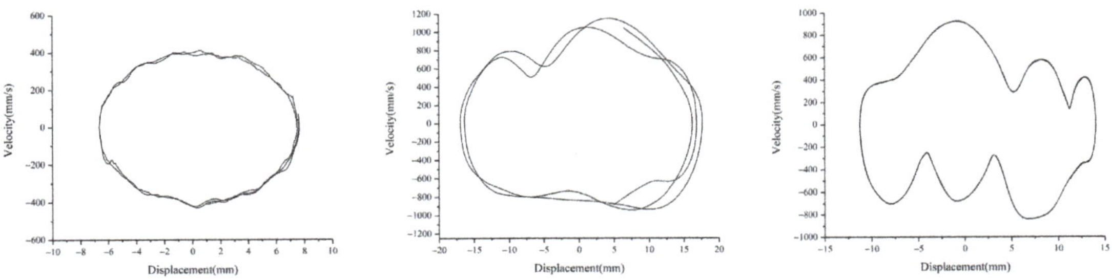

(a) (b) (c)

Figure 12. Experimental phase portrait of different magnetic distance: (**a**) $d = 22$ mm, $d_g = 8$ mm; (**b**) $d = 16$ mm, $d_g = 6$ mm; (**c**) $d = 16$ mm, $d_g = 8$ mm.

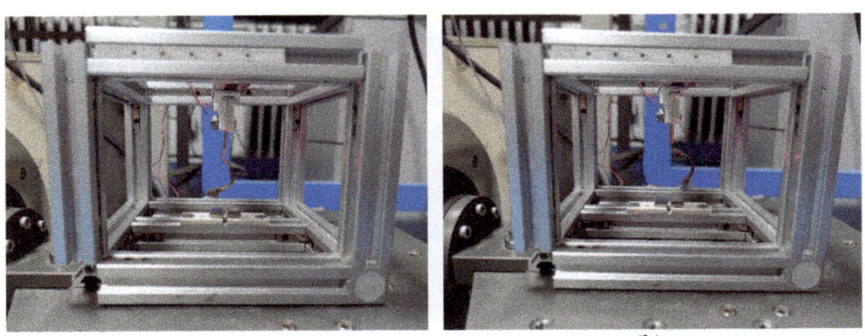

(a) (b)

Figure 13. Two stable equilibrium positions: (**a**) stable1; (**b**) stable2.

Comparing the experimental and simulation results, it can be seen that when the system response displacement is small, such as mono-stable motion, or the excitation is low, the experimental results are in good agreement with the simulation results, and the displacement amplitude error is about 8%. When the excitation is large or the system is in bi-stable and tri-stable motion, the experimental results shown in Figures 10c and 12b,c have slightly obvious errors, and the experimental phase diagram appears to be tilted. The main reasons are: (1) there are processing errors in the production of the linear-arch composite beam TPEH; (2) the phase diagram obtained by the experiment is asymmetric and inclined, while the phase diagram obtained by the simulation is not inclined and symmetrical. This is because the gravity factor is not considered in the simulation; (3) due to the existence of the arched part in the beam, there is a deviation in the data collected by the laser vibrometer.

5. Conclusions

In this paper, a dynamic model is established for the linear-arch composite beam TPEH. The dynamic equation is numerically solved by the fourth-order Runge–Kutta algorithm, and the influence of magnet spacing and excitation intensity on system dynamics is analyzed. The influence of characteristics is verified by experiments to verify the correctness of the theoretical analysis. The following main conclusions are obtained from simulation and experiment:

(1) When keeping dg unchanged, by changing d, the system can form a mono-stable system and a tri-stable system. When keeping d unchanged, increasing d_g, the system forms a bi-stable, tri-stable and mono-stable system in sequence. When the system moves in a tri-stable state, the vibration response displacement of the system increases significantly.

(2) Take the energy harvester with $d_g = 8$ mm and $d = 16$ mm as an example. Under this condition, the potential energy curve of the system has three potential wells, the depth of the potential wells is relatively shallow, and the width is relatively uniform; this helps the system to achieve a large response under low excitation.

(3) As the excitation level increases, it is easier for the system to cross the barrier to achieve inter-well movement, and the response displacement of the energy harvester increases.

(4) The asymmetric restoring force of the linear-arch beam results in an asymmetric potential well in the potential energy curve, which provides a new solution for the application of energy harvester in a low excitation environment.

Author Contributions: Conceptualization, X.Z. and L.C.; methodology, L.C.; software, L.C. and X.C.; validation, F.Z., Y.G. and L.C.; writing—original draft preparation, L.C.; writing—review and editing, X.C.; visualization, F.Z.; supervision, L.C.; project administration, X.Z.; funding acquisition, X.Z. All authors have read and agreed to the published version of the manuscript.

Funding: This research was funded by the National Natural Science Founds of China (Grant No. 51974228 and No. 51834006), the National Green Manufacturing System Integration Project (Grant No. 2017-327), Shaanxi Innovative Talent Plan Project (Grant No. 2018TD-032), and Key R&D project in Shaanxi (Grant No. 2018ZDCXL-GY-06-04).

Institutional Review Board Statement: Not applicable.

Informed Consent Statement: Not applicable.

Data Availability Statement: Not applicable.

Acknowledgments: Authors acknowledge the support from the students Lin Wang, Meng Zuo and Xiao She for their assistance in the modeling and experiments.

Conflicts of Interest: The authors declare no conflict of interest.

References

1. Li, K.; Xie, Z.; Zeng, D.; Li, W.; Luo, J.; Wang, M.; Ren, M.; Xia, C. Power cable vibration monitoring based on wireless distributed sensor network. *Procedia Comput. Sci.* **2021**, *183*, 401–411. [CrossRef]
2. Joris, L.; Dupont, F.; Laurent, P.; Bellier, P.; Stoukatch, S.; Redoute, J.-M. An Autonomous Sigfox Wireless Sensor Node for Environmental Monitoring. *IEEE Sensors Lett.* **2019**, *3*, 01–04. [CrossRef]
3. Tao, K.; Chen, Z.; Yi, H.; Zhang, R.; Shen, Q.; Wu, J.; Tang, L.; Fan, K.; Fu, Y.; Miao, J.; et al. Hierarchical Honeycomb-Structured Electret/Triboelectric Nanogenerator for Biomechanical and Morphing Wing Energy Harvesting. *Nano-Micro Lett.* **2021**, *13*, 1–16. [CrossRef] [PubMed]
4. Yang, Z.; Zhou, S.; Zu, J.; Inman, D. High-Performance Piezoelectric Energy Harvesters and Their Applications. *Joule* **2018**, *2*, 642–697. [CrossRef]
5. Sezer, N.; Ko, M. A Comprehensive Review on the State-of-the-Art of Piezoelectric Energy Harvesting. *Nano Energy* **2020**, *80*, 1–25.
6. Yu, L.; Tang, L.; Yang, T. Piezoelectric passive self-tuning energy harvester based on a beam-slider structure. *J. Sound Vib.* **2020**, *489*, 115689. [CrossRef]
7. Wang, B.; Li, Z.; Yang, Z. A distributed-parameter electromechanical coupling model for a piezoelectric energy harvester with variable curvature. *Smart Mater. Struct.* **2020**, *29*, 115015. [CrossRef]

8. Paknejad, A.; Rahimi, G.; Farrokhabadi, A.; Khatibi, M.M. Analytical solution of piezoelectric energy harvester patch for various thin multilayer composite beams. *Compos. Struct.* **2016**, *154*, 694–706. [CrossRef]
9. Wang, G.; Liao, W.-H.; Yang, B.; Wang, X.; Xu, W.; Li, X. Dynamic and energetic characteristics of a bistable piezoelectric vibration energy harvester with an elastic magnifier. *Mech. Syst. Signal Process.* **2018**, *105*, 427–446. [CrossRef]
10. Yildirim, T.; Ghayesh, M.H.; Li, W.; Alici, G. A review on performance enhancement techniques for ambient vibration energy harvesters. *Renew. Sustain. Energy Rev.* **2017**, *71*, 435–449. [CrossRef]
11. Tran, N.; Ghayesh, M.H.; Arjomandi, M. Ambient vibration energy harvesters: A review on nonlinear techniques for performance enhancement. *Int. J. Eng. Sci.* **2018**, *127*, 162–185. [CrossRef]
12. Li, P.; Liu, Y.; Wang, Y.; Luo, C.; Li, G.; Hu, J.; Liu, W.; Zhang, W. Low-frequency and wideband vibration energy harvester with flexible frame and interdigital structure. *AIP Adv.* **2015**, *5*, 047151. [CrossRef]
13. Liu, D.; Al-Haik, M.; Zakaria, M.; Hajj, M.R. Piezoelectric energy harvesting using L-shaped structures. *J. Intell. Mater. Syst. Struct.* **2018**, *29*, 1206–1215. [CrossRef]
14. Zhou, S.; Hobeck, J.D.; Cao, J.; Inman, D.J. Analytical and experimental investigation of flexible longitudinal zigzag structures for enhanced multi-directional energy harvesting. *Smart Mater. Struct.* **2017**, *26*, 35008. [CrossRef]
15. Wu, H.; Tang, L.; Yang, Y.; Soh, C.K. A novel two-degrees-of-freedom piezoelectric energy harvester. *J. Intell. Mater. Syst. Struct.* **2012**, *24*, 357–368. [CrossRef]
16. Rezaei, M.; Khadem, S.E.; Firoozy, P. Broadband and tunable PZT energy harvesting utilizing local nonlinearity and tip mass effects. *Int. J. Eng. Sci.* **2017**, *118*, 1–15. [CrossRef]
17. Liu, H.; Lee, C.; Kobayashi, T.; Tay, C.J.; Quan, C. Investigation of a MEMS piezoelectric energy harvester system with a frequency-widened-bandwidth mechanism introduced by mechanical stoppers. *Smart Mater. Struct.* **2012**, *21*, 035005. [CrossRef]
18. Kumar, T.; Kumar, R.; Chauhan, V.S.; Twiefel, J. Finite-Element Analysis of a Varying-Width Bistable Piezoelectric Energy Harvester. *Energy Technol.* **2015**, *3*, 1243–1249. [CrossRef]
19. Singh, K.A.; Kumar, R.; Weber, R.J. A broadband bistable piezoelectric energy harvester with nonlinear high-power extraction. *IEEET Power Electr.* **2015**, *30*, 6763–6774. [CrossRef]
20. Wang, G.; Ju, Y.; Liao, W.-H.; Zhao, Z.; Tan, J. A hybrid piezoelectric device combining a tri-stable energy harvester with an elastic base for low-orbit vibration energy harvesting enhancement. *Smart Mater. Struct.* **2021**, *30*, 075028. [CrossRef]
21. Zhou, Z.Y.; Qin, W.Y.; Zhu, P. Energy harvesting in a quad-stable harvester subjected to random excitation. *AIP Adv.* **2016**, *6*, 025022. [CrossRef]
22. Yao, M.; Liu, P.; Ma, L.; Wang, H.; Zhang, W. Experimental study on broadband bistable energy harvester with L-shaped piezoelectric cantilever beam. *Acta Mech. Sin.* **2020**, *36*, 557–577. [CrossRef]
23. Wang, G.; Wu, H.; Liao, W.-H.; Cui, S.; Zhao, Z.; Tan, J. A modified magnetic force model and experimental validation of a tri-stable piezoelectric energy harvester. *J. Intell. Mater. Syst. Struct.* **2020**, *31*, 967–979. [CrossRef]
24. Zhu, P.; Ren, X.; Qin, W.; Yang, Y.; Zhou, Z. Theoretical and experimental studies on the characteristics of a tri-stable piezoelectric harvester. *Arch. Appl. Mech.* **2017**, *87*, 1541–1554. [CrossRef]
25. Zhou, S.; Cao, J.; Inman, D.J.; Lin, J.; Liu, S.; Wang, Z. Broadband tri-stable energy harvester: Modeling and experiment verification. *Appl. Energ.* **2014**, *133*, 33–39. [CrossRef]
26. Zhang, X.; Zuo, M.; Yang, W.; Wan, X. A Tri-Stable Piezoelectric Vibration Energy Harvester for Composite Shape Beam: Nonlinear Modeling and Analysis. *Sensors* **2020**, *20*, 1370. [CrossRef]
27. Hai-Tao, L.; Hu, D.; Xing-Jian, J.; Wei-Yang, Q.; Li-Qun, C. Improving the performance of a tri-stable energy harvester with a staircase-shaped potential well. *Mech. Syst. Signal Process.* **2021**, *159*, 107805. [CrossRef]
28. Wang, G.; Zhao, Z.; Liao, W.-H.; Tan, J.; Ju, Y.; Li, Y. Characteristics of a tri-stable piezoelectric vibration energy harvester by considering geometric nonlinearity and gravitation effects. *Mech. Syst. Signal Process.* **2020**, *138*, 106571. [CrossRef]
29. Sun, S.; Leng, Y.; Su, X.; Zhang, Y.; Chen, X.; Xu, J. Performance of a novel dual-magnet tri-stable piezoelectric energy harvester subjected to random excitation. *Energy Convers. Manag.* **2021**, *239*, 114246. [CrossRef]
30. Zhang, X.; Yang, W.; Zuo, M.; Tan, H.; Fan, H.; Mao, Q.; Wan, X. An Arc-shaped Piezoelectric Bistable Vibration Energy Harvester: Modeling and Experiments. *Sensors* **2018**, *18*, 4472. [CrossRef] [PubMed]
31. Erturk, A.; Inman, D.J. *Piezoelectric Energy Harvesting*; John Wiley & Sons: Hoboken, NJ, USA, 2011.
32. Zhu, P.; Ren, X.; Qin, W.; Zhou, Z. Improving energy harvesting in a tri-stable piezomagnetoelastic beam with two attractive external magnets subjected to random excitation. *Arch. Appl. Mech.* **2017**, *87*, 45–57. [CrossRef]
33. Wang, W.; Cao, J.; Bowen, C.R.; Litak, G. Multiple solutions of asymmetric potential bistable energy harvesters: Numerical simulation and experimental validation. *Eur. Phys. J. B* **2018**, *91*, 254. [CrossRef]

micromachines

MDPI

Article

A Curve-Shaped Beam Bistable Piezoelectric Energy Harvester with Variable Potential Well: Modeling and Numerical Simulation

Xiaoyu Chen [1,2], Xuhui Zhang [1,3,*], Luyang Chen [1], Yan Guo [1] and Fulin Zhu [1]

1 College of Mechanical Engineering, Xi'an University of Science and Technology, Xi'an 710054, China; 19105016004@stu.xust.edu.cn (X.C.); chenluyang@stu.xust.edu.cn (L.C.); g1014901143@163.com (Y.G.); 20205224060@stu.xust.edu.cn (F.Z.)
2 College of Engineering, Zunyi Normal College, Zunyi 563006, China
3 Shaanxi Key Laboratory of Mine Electromechanical Equipment Intelligent Monitoring, Xi'an 710054, China
* Correspondence: zhangxh@xust.edu.cn; Tel.: +86-133-6397-5328

Abstract: To improve the energy harvesting performance of an energy harvester, a novel bistable piezoelectric energy harvester with variable potential well (BPEH-V) is proposed by introducing a spring to the external magnet from a curve-shaped beam bistable harvester (CBH-C). First, finite element simulation was performed in COMSOL software to validate that the curved beam configuration was superior to the straight beam in power generation performance, which benefits energy harvesting. Moreover, the nonlinear magnetic model was obtained by using the magnetic dipoles method, and the nonlinear restoring force model of the curve-shaped beam was acquired based on fitting the experimental data. The corresponding coupled governing equations were derived by using generalized Hamilton's principle, the dynamic responses were obtained by solving the coupling equations with the ode45 method. Finally, the numerical simulations showed that the proposed harvester can make interwell oscillations easier due to the spring being efficiently introduced to pull down the potential barrier compared with the conventional bistable harvester. Spring stiffness has a great impact on characteristics of the system, and a suitable stiffness contributes to realize large-amplitude interwell oscillations over a wide range of excitation, especially in the low excitation condition.

Keywords: energy harvesting; curve-shaped configuration; variable potential well; dynamic behavior

Citation: Chen, X.; Zhang, X.; Chen, L.; Guo, Y.; Zhu, F. A Curve-Shaped Beam Bistable Piezoelectric Energy Harvester with Variable Potential Well: Modeling and Numerical Simulation. *Micromachines* **2021**, *12*, 995. https://doi.org/10.3390/mi12080995

Academic Editors: Kai Tao and Yunjia Li

Received: 24 July 2021
Accepted: 16 August 2021
Published: 21 August 2021

1. Introduction

Wireless sensor, wearable devices, and medical implants have shown their significance in modern society [1]. Powering these low-power devices is usually done through conventional batteries, however, these batteries must be regularly recharged or replaced, which can be very costly and cumbersome [2]. Meanwhile, there are environmental issues when disposing of used batteries after operation [3]. Energy harvesting technology holds great potential to achieve the self-powered operation of these devices. Among the various energy harvesting technologies, electromagnetism, electrostatics, and piezoelectricity are the three main methods that generate energy from vibration [4,5]. In particular, vibration-based piezoelectric energy technology can convert kinetic energy from the ambient environment via piezoelectric effect to electric energy, which has received considerable interest for its high energy density, ease of implementation, and miniaturization [6].

At the early stage, research on piezoelectric energy harvesters was mainly based on a linear piezoelectric energy harvester. The linear piezoelectric energy harvester has a high resonance frequency, and when the environmental frequency deviates from its resonance frequency, the power generation performance of the system will drop sharply, resulting in low environmental adaptability [7]. Currently, nonlinear bistable piezoelectric energy harvesters have received great attention. Zhang et al. [8] proposed an arched composite beam magnetically coupled piezoelectric energy harvester. Experiments showed that the effective bandwidth of the energy harvester under magnetic coupling was 3.1 times the bandwidth without magnetic force. Rubes et al. [9] conducted research on magnetically coupled bistable piezoelectric energy harvesters, and their research showed that the introduction of nonlinear stiffness can greatly improve the energy harvesting performance of piezoelectric energy harvesters; Erturk and Inman [10] experimentally proved that the nonlinearity of magnetic coupling can cause vibration between the bistable high-energy traps, thereby improving the collection performance of the energy harvester; and Stanton et al. [11] established a complete dynamic model for the output voltage and dynamic behavior of the magnetic coupling bistable piezoelectric energy harvester and proved the availability of the bistable harvester. Under the condition of simple harmonic excitation, Li et al. [12] developed a magnetic-coupled bi-stable flutter-based energy harvester and proved that the proposed system was an effective design approach for enhancing energy harvesting capability in a low air speed range. Singh et al. [13] investigated a bistable piezoelectric energy harvester with SSHI circuit, and their experiments proved that the output power of the bistable piezoelectric energy harvester with the SSHI circuit reached 478 μw, while the corresponding linear structure was only 129 μw.

The above research shows that the bistable piezoelectric energy harvester is effective for improving the performance of an energy harvester. However, the above harvesters all had a fixed barrier height. In practical applications, the excitation level must provide enough energy to overcome the barrier to achieve a large response, otherwise it will not be able to work well, resulting in poor output performance. In order to reduce the barrier height to improve the performance of the bistable piezoelectric energy harvester, many scholars have carried out studies on piezoelectric energy harvesters with variable potential wells. Zhou et al. [14] placed an external magnet in the middle of the fixed beams at both ends and proposed a bistable system with variable potential wells. Experiments proved that the system not only had a low interwell jump threshold, but also produced higher voltage output. Cao et al. [15] proposed a bistable energy harvesting with time varying potential energy to harvest energy from human motion and various motion speed treadmill tests were performed to demonstrate the advantage of time-varying bistable harvesters over linear and monostable. Nguyen et al. [16] proposed a bistable piezoelectric energy harvester with an auxiliary magnet oscillator and their research showed that this design could improve 114–545% bandwidth compared with traditional bistable piezoelectric energy harvesters. Yang et al. [17–19] designed a double-beam piezoelectric energy harvester with variable potential well structure and verified its advantages over traditional bistable piezoelectric energy harvester under random excitation conditions. Lan et al. [20] significantly reduced the barrier height of the traditional bistable piezoelectric energy harvester by adding a small magnet to a traditional bistable energy harvester and compared their design with a three-stable piezoelectric energy harvester, verifying the validity of the proposed device. Shan et al. [21] designed an elastically connected bistable piezoelectric energy harvester based on the straight beam configuration, where the energy harvester had a variable potential barrier during the vibration process. It was experimentally proven that the energy harvesting bandwidth was 60% higher than that of the traditional energy harvester. Li et al. [22] carried out theoretical analysis on the elastically connected straight beam piezoelectric energy harvester, and the results showed that the spring-connected bistable piezoelectric energy harvester had a variable potential function and better energy harvesting performance under low-frequency excitation. Kim et al. [23] designed a multi-degree of freedom (MDOF) vibration energy harvesting system that leverages magnetically coupled bistable and linear harvesters, where the analytical, numerical, and experimental investigations revealed that the novel harvester could facilitate the energetic interwell response for relatively low excitation amplitudes and frequencies by passively and adaptively lowering the potential energy barrier level. Qian et al. [24] developed a broadband piezoelectric energy harvester (PEH) with a mechanically tunable potential function, and the simulations proved that the proposed PEH could harvest vibration energy in a wide frequency range of 0–91 Hz at the excitation level of 0.5 g.

Inspired by the development of variable-potential-energy techniques, this paper proposes a novel bistable energy harvester with a variable potential well. Meanwhile, we used a curve-shaped beam as the energy transducing element to further improve the performance of the piezoelectric energy harvester due to the disadvantages of the straight beam in terms of uneven stress, low conversion efficiency [25,26]. The finite element simulation was performed for the curve-shaped beam and the conventional beam. The results show that the curved beam structure has a special stress distribution and can improve output voltage compared with the straight beam structure. Then, the dynamic model of BPEH-V system is established. Numerical simulation analysis showed that it was easier for the proposed harvester to achieve large-amplitude response in a low-frequency environment compared with the conventional counterpart, and the spring stiffness had an important impact on system performance. The research can provide theoretical guidance for the optimal design and engineering application of the novel piezoelectric energy harvester.

2. Finite-Element Simulation

2.1. Stress Analysis

At present, most of these existing piezoelectric energy harvesters utilize straight beam as the energy transducing elements due to its advantages in terms of simplicity and ease of fabrication, as shown in Figure 1a.

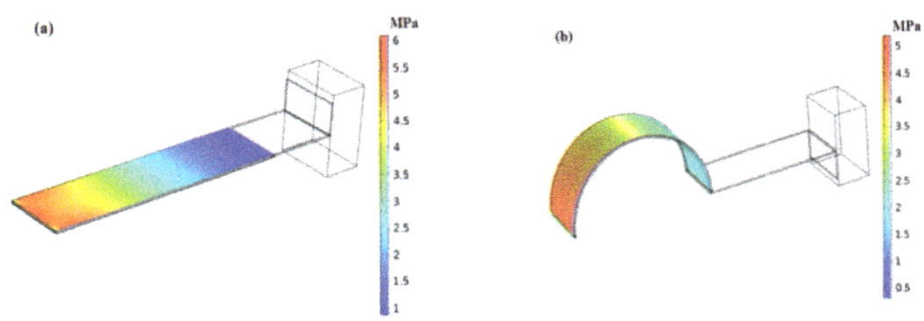

Figure 1. Finite-element model: (**a**) Straight beam, (**b**) Curve-shaped beam.

As we know, the conversion efficiency of piezoelectric materials is closely related to the stress distribution of the base layer. The evenly-distributed stress is helpful for harvesting energy and improving conversion efficiency. According to the theory of material mechanics, the conventional straight cantilever experiences a linear stress distribution on the surface when excited. The base layer considered in this work is schematically shown in Figure 1b, which is built of an arc-shaped and a flat configuration, and experiences different stress distribution from the conventional straight cantilever due to the arc-shaped configuration being introduced to improve the stress condition. The finite element analysis was performed in COMSOL software to analyze the influence of curved beam and traditional straight beam structure on the stress distribution of piezoelectric materials (PVDF). In order to make a fair comparison, both beams had the same rectangular sections; the material parameters used are listed in Table 1. Two identical mass were attached at the free end of both beams to reduce resonance frequency, respectively. Note that the curve-shaped beam had an arch with a central angle of 180 degrees, with a radius of R = 10 mm. The PVDF was only adhered to the arc-shaped surface of the curve-shaped beam, with a horizontal length of Lp = 31.4 mm. Meanwhile, the identical piezoelectric material (PVDF) was attached on the surface of the straight beam. The same load was applied on both beams, respectively. The stress distribution along the length direction of the piezoelectric materials on the curved beam and straight beam is shown in Figure 2, respectively.

Table 1. Material parameters for simulation.

Parameter	Symbol	Value	Unit
	Substrate layer (beryllium bronze)		
Density	ρ_s	8300	kg/m^3
Elastic modulus	E_s	128	GPa
Arc-shaped radius	R_s	10×10^{-3}	m
Horizontal length	L_s	40×10^{-3}	m
Height	h_s	2×10^{-4}	m
Width	b_s	8×10^{-3}	m
	Piezoelectric layer (PVDF)		
Density	ρ_p	1780	kg/m^3
Elastic modulus	E_p	3	GPa
Length	L_p	31.4×10^{-3}	m
Height	h_p	1.1×10^{-4}	m
Width	b_p	8×10^{-3}	m
Piezoelectric stress constant	$\bar{e}_{31}$	11.5	C/m^2

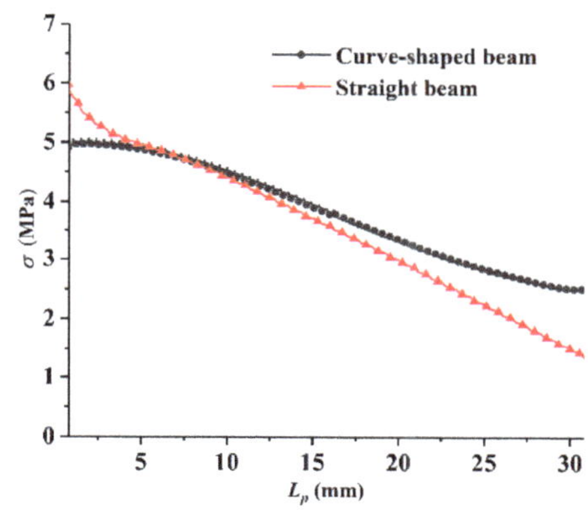

Figure 2. Comparison of stress distribution of PVDF on the curve-shaped beam and the straight beam.

It can be seen from Figure 2 that the stress of the piezoelectric material on the straight beam structure decreased linearly from the fixed end. The stress of the piezoelectric material on the curved beam structure was higher than that of the straight beam structure in most areas, and dropped more smoothly than that of the curved beam. The stress distribution was correlated with the bending moment acting on the configuration, the bending moment acting on the straight beam configuration decreased linearly along the fixed end, leading to linearly decreasing stress. However, the bending moment acting on the arc-shaped configuration behaved in a complex manner and decreased nonlinearly along the fixed end according to the theory of material mechanics, thus improving the stress distribution.

2.2. Generation Performance Comparisons

The piezoelectric coupling analyses are carried out in COMSOL software to compare the power generation performance of the curved beam and the straight beam structure.

Figure 3 shows the voltage comparison diagram of the curved beam and the straight beam structure under two different excitation conditions. At the excitation level of 2 m/s^2, the resonance voltage of the curved beam was 11 V, and the resonance voltage of the straight beam was 7 V. With an increase in the excitation level to 5 m/s^2, the resonance voltage of the curved beam was 22 V, and corresponding value of the straight beam was only 15 V in this case. Based on the simulation results, the voltage output of the piezoelectric material on the curved beam structure is always higher than that of the straight beam structure under two different excitation levels. The relatively large and evenly-distributed stress results in less energy dissipation during charge flowing from the large stress region to low, which contributes to enhance the power output and energy conversion efficiency [27,28]. The special stress distribution of the curved beam configuration is beneficial to improving the output performance of the piezoelectric material. Therefore, the piezoelectric material on the surface of the curved beam produces a higher output voltage than that of the straight beam, and the curved beam has a better performance than the straight beam. At the same time, it can be found that curved beams have a lower resonance frequency than the straight beam, which will also benefit energy harvesting in low-frequency environments. Therefore, the introduction of a curved beam structure to a piezoelectric energy harvester is beneficial to increase the output power and improve the output performance of the conventional energy harvester.

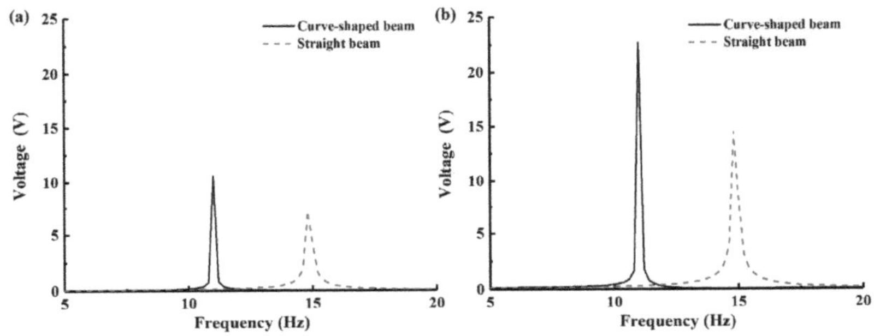

Figure 3. Output voltage obtained by finite-element simulation at different excitation amplitude: (**a**) $A = 2\ \mathrm{m/s^2}$; (**b**) $A = 5\ \mathrm{m/s^2}$.

3. BPEH-V Configuration

The BPEH-V, shown in Figure 4, is comprised of a curve-shaped beam, magnet A, magnet B (i.e., external magnet), piezoelectric material (PVDF), and base. The piezoelectric material is attached to the surface of the arched part of the curve-shaped beam to realize energy conversion, and the flat part remains free. The external magnet B maintains a magnetic repulsive relationship with magnet A, and imposing bistability on the system. The difference between the proposed system and a conventional bistable piezoelectric harvester is because the external magnet B is connected to the base through a spring. If the BPEH-V is excited by ambient vibrations, the piezoelectric cantilever and magnet A are vibrated with the base, so the oscillation of piezoelectric cantilever would result in the deformation of PVDF, thus the conversion of mechanical energy from ambience into electrical energy via the piezoelectric effect can be achieved. When the end magnet of the cantilever beam moves to the intermediate equilibrium position, the spring is compressed and the potential barrier is lowered. Conversely, if the end magnet moves far away from the intermediate equilibrium position, the spring returns to the zero point and the magnetic distance is reduced to maintain the bistable characteristics of the system. Therefore, a bistable piezoelectric energy harvester with variable potential well is formed during the process of the piezoelectric beam vibration.

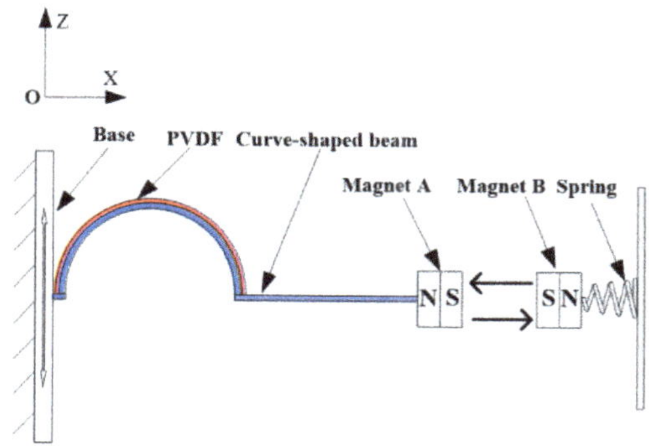

Figure 4. Schematic diagram of the BPEH-V.

The BPEH-V not only retains the vibration bistability of the piezoelectric cantilever but could also adjust the potential barrier level, which is beneficial to realizing large-amplitude interwell oscillations under a low excitation level, thus improving the energy harvesting performance.

3.1. Theoretical Modeling

3.1.1. Modeling of Nonlinear Restoring Force

Unlike the linear restoring force of the conventional straight beam, the restoring force was nonlinear in the curve-shaped beam due to the existence of the arc-shaped configuration. To model the restoring force, the relationship between deflection and restoring force is extracted by using experimental method. To this end, the curve-shaped beam was fixed on the left end, and the free end of the beam was pushed by the dynamometer to measure the value of the nonlinear restoring force at different displacements. The process was repeated and the measurement results were averaged, then the relationship between the resorting force and transverse displacements were fit to a polynomial, as follows:

$$F_r = k_1 u^3(L,t) + k_2 u^2(L,t) + k_3 u(L,t) \tag{1}$$

where k_1, k_2, and k_3 are constant coefficients on the third, second, and first-order terms, respectively. Figure 5 shows the measurement results and curve fitting results of the nonlinear restoring force of the curve-shaped beam. It can be observed from Figure 5 that the experimental data and the fitting curve had good agreement, and the restoring force of the curve-shaped beam exhibited a curve due to the existence of the curved configuration. Setting $u = 0$ as the static equilibrium position, it was found that the measurement results were asymmetrical, which is due to the fact that the radius of curvature for the curved configuration is continuously varied in the process of the piezoelectric beam vibration, and resulting in asymmetric nonlinear restoring force.

3.1.2. Modeling of Magnetic Force

The permanent magnets can be modeled as the point dipoles when calculating the magnetic force between the tip magnet and the external magnet. The schematic diagram of the spatial position of the magnets is shown in Figure 6. Considering the additional degree of freedom (DOF) and rotation of the magnet, the distance vector r_{BA} from the center of magnet B to magnet A can be expressed as:

$$r_{BA} = [-d - q(t),\ u(L,t)] \tag{2}$$

where $q(t)$ is the compression displacement of magnet B, and the magnetic field generated by magnet B on magnet A is obtained as [29]:

$$U_{MA} = \frac{\mu_0}{4\pi}\left[\frac{m_B}{\|r_{BA}\|_2^3} - \frac{(m_B \cdot r_{BA}) \cdot 3 \cdot r_{BA}}{\|r_{BA}\|_2^5}\right] m_A \tag{3}$$

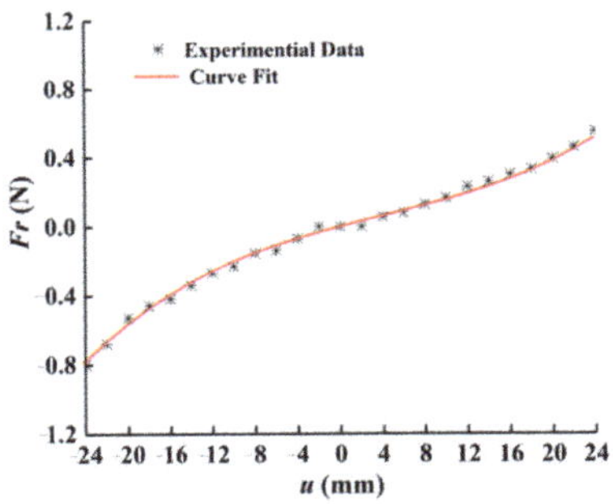

Figure 5. Displacement-restoring force curve of the curve-shaped beam.

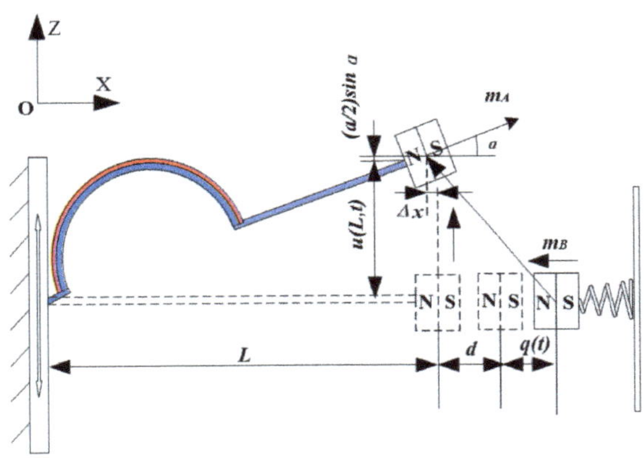

Figure 6. Schematic diagram of the spatial position of the magnets.

The magnetic moment vectors m_A and m_B for magnets A and B can be respectively expressed as:

$$m_A = [M_A V_A cos\alpha,\ M_A V_A sin\alpha] \tag{4}$$

$$m_B = [-M_B V_B,\ 0] \tag{5}$$

where M_i and V_i (i = A, B) are the magnetization strength and material volume of magnets A and B, respectively. α is the slope of beam at the free end, which is given by:

$$\alpha = arctan(\dot{u}(L,t)) \tag{6}$$

Substituting Equation (2) and Equation (4) to Equation (6) into Equation (3), the magnetic field U_{MA} can be expressed in the following equation:

$$U_{MA} = \frac{\mu_0 M_A V_A M_B V_B(-u(L,t)^2 - 2(d+q(t))^2 + 3(d+q(t))u(L,t)\dot{u}(L,t)))}{4\pi\sqrt{(\dot{u}(L,t))^2+1}\,(u(L,t)^2+(d+q(t))^2)^{5/2}} \tag{7}$$

3.1.3. Dynamical Model

To predict the response of BPEH-V, considering the Euler–Bernoulli theory and the linear constitutive equations for piezoelectric materials, the coupled governing equations are derived by using the generalized Hamilton principle.

$$\int_{t1}^{t2} [\delta(T_k - U_r + W_e) + \delta W_{nc} - \delta U_m] = 0 \tag{8}$$

where U_r is the elastic potential energy of the piezoelectric beam and U_m is the magnetic potential energy. W_e is the electric potential energy of the piezoelectric layer, and W_{nc} is the external work applied to the system. The whole kinetic energy of the proposed system can be expressed as:

$$T_k = T_1 + T_2 + T_3 + T_4 \tag{9}$$

where T_1, T_2, T_3, and T_4 represent the kinetic energy of the substrate layer, the piezoelectric layer, the tip magnet A, and the movable magnet B.

$$T_k = \frac{1}{2}\int\limits_{V_s} \rho_s \dot{u}^2(x,t)dV_s + \frac{1}{2}\int\limits_{V_p} \rho_p \dot{u}^2(x,t)dV_p + \frac{1}{2}m_A \dot{u}^2(x,t)|_{x=L} \ldots + \frac{1}{2}I_t\left[\frac{\partial^2 u(x,t)}{\partial t \partial x}|_{x=L}\right]^2 + \frac{1}{2}m_B\dot{q}(t)^2 \tag{10}$$

where $u(x,t)$ is the transverse displacement of the beam; V_p and V_s are the piezoelectric and substrate layer volume, respectively; and I_t is the rotational inertia of the tip magnet with respect to the beam free end. The electric potential energy of the piezoelectric material can be expressed as follows:

$$W_e = \frac{1}{2}\int_{V_p} \varepsilon_{33}^s E_3^2 V_p + \frac{1}{2}\int_{V_p} e_{31} E_3 S_1 dV_p \tag{11}$$

where E_3 and S_1 represent the electrical field and the axial strain, respectively. ε_{33}^s and e_{31} represent the permittivity component at constant strain and the piezoelectric constant. The external work applied to the BPEH-V system can be written as follows:

$$\delta W_{nc} = -\int_0^L \delta u(x,t)m(x)\ddot{z}(t)dx - \delta u(L,t)m_0\ddot{z}(t) + Q\delta v \tag{12}$$

In this paper, based on the Rayleigh–Ritz principle, it is assumed that a single-mode approximation of the beam deformation is sufficient, and the vibrational displacement of the beam can be expressed as follows:

$$u(x,t) = \sum_{i=1}^{n} \varphi_i(x) r_i(t) \tag{13}$$

where $\varphi_i(x)$ is the ith mode shape of the beam and $r_i(t)$ is the time-dependent generalized coordinates. Under the low frequency excitations, the vibration of the beam is mainly concentrated in the first-order mode, so it is sufficient to consider one mode to obtain the reduced-order model. Meanwhile, for the boundary conditions where one end is clamped and the other one is free, the allowable function can be written as [30,31]:

$$\varphi(x) = 1 - \cos\left(\frac{\pi x}{2L}\right) \tag{14}$$

Substituting Equations (7) and (10)–(12) into Equation (8), according to Kirchhoff's law, the governing equations of BPEH-V system are obtained:

$$M\ddot{r}(t) + C\dot{r}(t) + k_1 r^3(t) + k_2 r^2(t) + k_3 r(t) - \frac{\partial U_{MA}}{\partial r(t)} - \theta v = -H_s \ddot{z}(t) \tag{15}$$

$$\theta \dot{r}(t) + C_p \dot{v}(t) + \frac{v(t)}{R} = 0 \tag{16}$$

$$m\ddot{q}(t) + kq(t) - F_q = 0 \tag{17}$$

where M and C refer to the mass coefficient and the damping coefficient, respectively. θ is the electromechanical coupling coefficient; C_p is the capacitance of the piezoelectric patch; R is the load resistance; and F_q is the horizontal magnetic force component, as follows:

$$M = \int_{\Omega b} \rho_b \varphi(x)^2 d_{\Omega b} + \int_{\Omega p} \rho_p \varphi(x)^2 d_{\Omega p} + m_0 \varphi(L)^2 + I_t \varphi'^2(L) \tag{18}$$

$$H_s = \rho_b A_b \int_0^L \varphi(x)dx + \rho_p A_p \int_0^{L_p} \varphi(x)dx + m_0 \tag{19}$$

$$C = \int_0^L c\varphi(x)^2 dx \tag{20}$$

$$\theta = \frac{1}{h_p} \int_{\Omega p} e_{31} z \varphi''(x) d_{\Omega p} \tag{21}$$

$$C_p = \frac{\varepsilon_{33}^S b_p L_p}{h_p} \tag{22}$$

$$F_q = \frac{\partial U_{MA}}{\partial q(t)} \tag{23}$$

4. Numerical Simulation

4.1. Study on the Potential Energy of BPEH-V

Magnetic potential energy is an important factor that affects the nonlinearity of the system. Different magnetic distances will produce different nonlinear magnetic forces, so the system presents different characteristics. Regarding the BPEH-V system, the magnetic potential energy is continuously varied with vibration due to the external magnet being connected elastically. Figure 7 shows the potential energy curve of the system under the condition of magnetic distance (d = 17 mm). In this case, two obvious potential wells are formed, that is, the system becomes bistable. We should notice that magnetic distance d is constantly varied during the vibration of the piezoelectric cantilever beam, so the potential energy of the system is different from a traditional bistable piezoelectric energy harvester with a fixed external magnet. The magnetic potential energy is not only affected by the magnetic distance d, but also by the compression displacement $q(t)$ of the spring. As shown in Figure 7, the x-axis denotes displacement of the curved-shape beam's tip, the y-axis denotes the compression displacement of the spring, and the z-axis denotes the potential energy of the system. The height of the barrier between the two wells is pulled down as the compression displacement of the spring gradually increases due to the repulsive force between the tip magnet and the external magnet. In this condition, the system can cross the potential barrier to realize interwell oscillations more easily. The influence of $q(t)$ in BPEH-V, which is caused by spring compression, equals that of time-varying d in the traditional bistable system. When the tip magnet tends to approach its original point (at $u(L,t) = 0$ in Figure 6), it drives the external magnet away from the equilibrium position due to magnetic repulsion, thus decreasing the potential barrier. Conversely, when the tip magnet moves far away from the original point, the potential barrier gradually becomes high and reaches its maximum. Thus, the design of the BPEH-V provides an adaptive potential using the spring in comparison to the traditional bistable system.

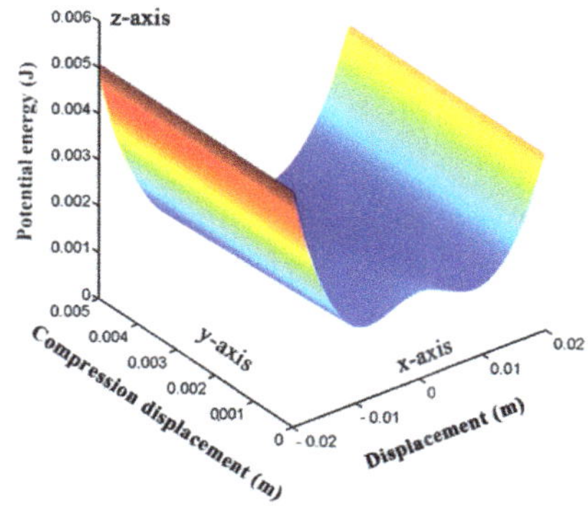

Figure 7. Potential curve of the system at different compression displacement of the spring.

Meanwhile, it can also be seen from Figure 7 that the potential energy curve of the proposed system is inconsistent with the straight beam bistable piezoelectric energy harvester. The potential well is shallower on the left side and deeper on the right side, showing an asymmetrical trend. This is mainly due to the asymmetric restoring force of the curve-shaped beam.

4.2. The Dynamics Analysis of BPEH-V

According to the potential energy diagram shown in Figure 7, the system becomes bistable and the height of the potential barrier is relatively shallow when the magnet distance is d = 17 mm. In this section, the numerical simulations are performed for the separation distance d = 17 mm to investigate the influence of the variable potential well on the dynamic characteristics of BPEH-V (the ode45 command of MATLAB was used here).

The bifurcation diagram of the tip displacement versus the excitation frequency of the BPEH-V and the CBH-C for excitation amplitude A = 10 m/s^2 is shown in Figure 8. Compared to Figure 8a,b, it can be found that BPEH-V exhibited more complex dynamic behaviors than the CBH-C. At 4 Hz excitation, BPEH-V enters into the chaotic oscillation, which can be concluded from the phase plane portrait (the phase plane portrait is drawn by red curves) and Poincaré map (the Poincaré map is drawn by black dots) depicted in Figure 9a.

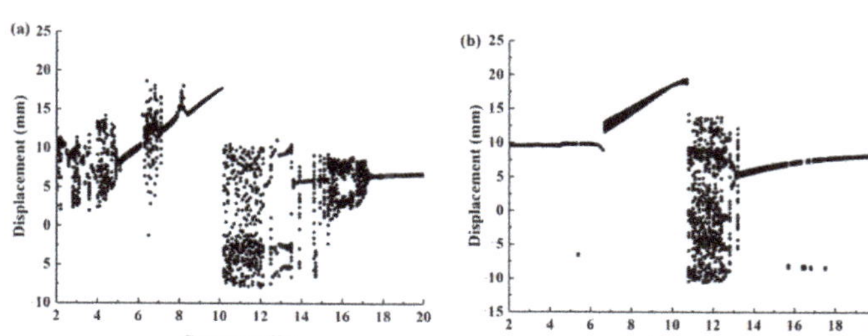

Figure 8. Bifurcation diagram of the tip displacement versus the excitation frequency for excitation amplitude A = 10 m/s^2: (**a**) BPEH-V; (**b**) CBH-C.

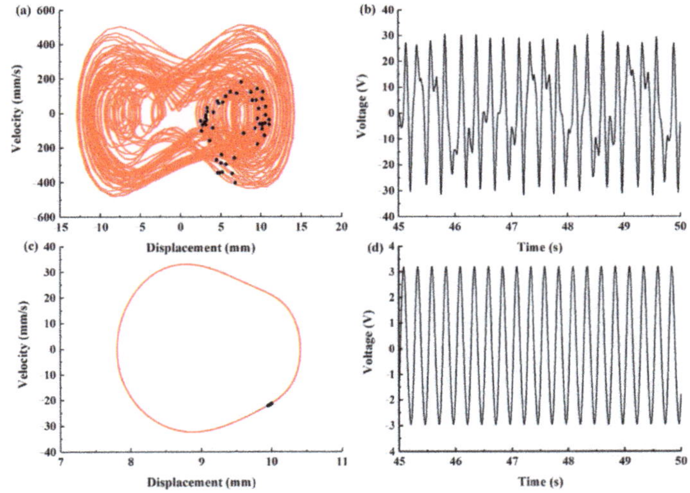

Figure 9. Phase plane portrait, Poincaré map, and output voltage histories for excitation frequency $f = 4$ Hz. (**a**) Phase plane portrait and Poincaré map. (**b**) Voltage histories for BPEH-V, respectively. (**c**) Phase plane portrait and Poincaré map. (**d**) Voltage histories for CBH-C, respectively.

However, the CBH-C system only made a small-amplitude intrawell motion at this time, as shown in Figure 9c. Meanwhile, compared with Figure 9b,d, we found that the BPEH-V generated a much higher output voltage than the CBH-C in the low excitation frequency.

With the increase in excitation frequency to 5 Hz (Figure 10 shows the simulation results for BPEH-V), the Poincaré map is concentrated in a single point and the phase plane portrait consists of a closed obit, as shown in Figure 10a, which demonstrates that the BPEH-V entered into large-amplitude periodic oscillations. However, the CBH-C system still made a small-amplitude intrawell motion at this time, as can be found from the bifurcation diagram of the tip displacement versus the excitation frequency depicted in Figure 8b.

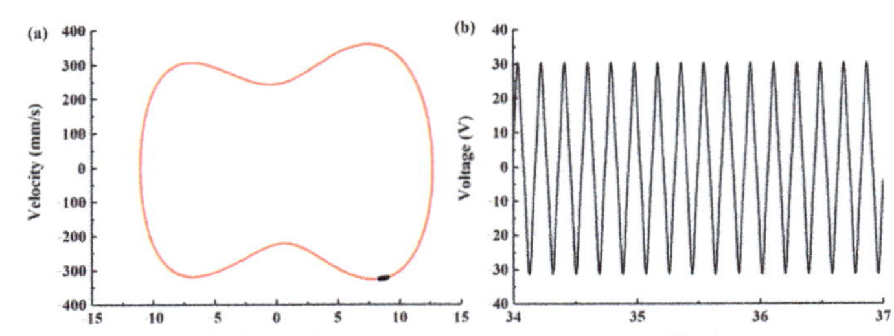

Figure 10. Simulation results for BPEH-V under excitation frequency $f = 5$ Hz. (**a**) Phase plane portrait and Poincaré map. (**b**) Output voltage histories.

With the increase in excitation frequency to 7.1 Hz, Figure 11 shows the simulation results for CBH-C, where the Poincaré map is concentrated in a single point and the phase plane portrait consists of a closed obit, which demonstrates that the CBH-C enters into large-amplitude periodic oscillations. Meanwhile, it was observed from Figure 8a that the BPEH-V underwent transient chaotic oscillation at 6.3 Hz excitation, and then returned to large-amplitude interwell oscillations at 7 Hz excitation.

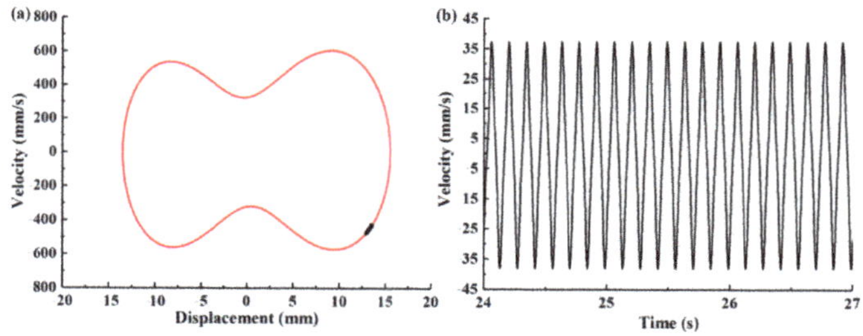

Figure 11. Simulation results for CBH-C under excitation frequency f = 7.1 Hz. (**a**) Phase plane portrait and Poincaré map. (**b**) Output voltage histories.

With the still further increase in excitation frequency, the BPEH-V exits large-amplitude interwell oscillations when the excitation frequency exceeds 10.1 Hz. Meanwhile, the CBH-C exits large-amplitude interwell oscillations at a frequency f = 10.3 Hz.

From the above simulations and analyses, we found that the frequency ranges of large-amplitude periodic response of BPEH-V were $5 < f < 6.3$ Hz and $7 < f < 10.1$ Hz, and the effective bandwidth was 4.4 Hz. The corresponding frequency range of CBH-C was $7.1 < f < 10.3$ Hz, and the effective bandwidth was only 3.2 Hz. Accordingly, the effective bandwidth of BPEH-V was 1.37 times that of CBH-C under the same circumstances due to the spring being efficiently introduced to broaden bandwidth, and the BPEH-V was superior to the CBH-C from the aspect of effective bandwidth. The conventional bistable system only made a small-amplitude intrawell motion at low excitation frequency due to the lack of sufficient energy to overcome the potential barrier. However, thanks to the compression adjustment of the spring, it can pull down the potential barrier and form an adaptive potential barrier. The BPEH-V with suitable stiffness can realize large-amplitude interwell motions at the lower excitation frequency, thus improving the harvesting performance.

4.3. The Influence of the Spring Stiffness K on Harvesting Performance

Spring stiffness has a great impact on the system characteristics. In order to investigate the influence of the spring stiffness K on energy harvesting performance, the numerical frequency-swept experiments of the BPEH-V system with three distinct spring stiffness were conducted under the excitation amplitude of 5 m/s^2, as shown in Figure 12. The BPEH-V system with suitable stiffness of K = 200 N/m can realize large-amplitude interwell oscillations and have a higher output at a frequency range of f = 9.4–11Hz. We decreased the spring stiffness to K = 150 N/m. The spring was more easily compressed due to the small spring stiffness, so the system could realize large-amplitude interwell oscillations at lower excitation frequency; the theoretical frequency range of the large-amplitude periodic response was f = 8.6–11 Hz; and the effective bandwidth was 2.4 Hz, which was broader than the case of K = 200 N/m. Meanwhile, we notice that the system without spring can only realized intrawell oscillations and generated a lower output voltage at the same condition, which was because that the system could not obtain sufficient energy at low excitation level to overcome the potential barrier, thus resulting in poor output performance. Therefore, the BPEH-V with suitable spring stiffness contributed to realize large-amplitude interwell oscillations over a wide range of excitation, especially in low excitation level compared to CBH-C.

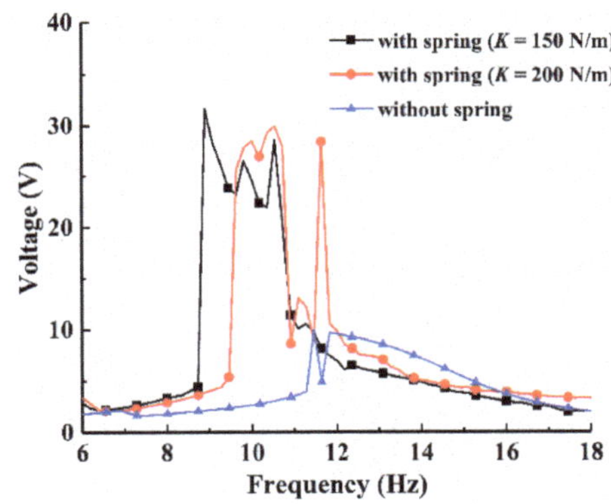

Figure 12. Frequency-swept voltage response of the system with different spring stiffness under excitation amplitude $A = 5\ \text{m/s}^2$.

It needs to be mentioned that we should ensure the bistable characteristic of the system when choosing a small stiffness spring as the connection element. Otherwise, the system will be close to a linear one and vibrates around the middle equilibrium point, leading to poor output performance.

5. Conclusions

This paper proposed a magnetically coupled bistable piezoelectric energy harvester based on an elastically connected external magnet. First, finite-element simulations were performed for the curve-shaped composite and the straight beam to compare the influence of different configurations on the stress distribution and power generation performance. Moreover, the dynamics model of the system was established by using the generalized Hamilton variational principle, and the fourth-order Runge–Kutta algorithm was used to numerically solve the dynamic equations. The dynamic characteristics of the piezoelectric energy harvester were analyzed and compared with the traditional curve-shaped beam bistable harvester. Finally, the influence of the spring stiffness on energy harvesting performance of the system was discussed. The main conclusions are as follows:

1. The curve-shaped configuration beams had a larger and more uniform strain distribution than the straight beam due to the special arched structure. Under the same excitation conditions, compared with the traditional straight beam, the curve-shaped configuration beam had a higher output voltage. Therefore, the curve-shaped beam was introduced into the nonlinear piezoelectric energy harvester, which can help to improve the harvesting efficiency of the energy harvesting device.
2. A spring was connected with an external magnet to form an elastically supported bistable system. The potential energy of the system was affected by the magnetic distance and spring stiffness. The elastic connection of the external magnet could adjust the height of the system's barrier to realize an adaptive potential barrier. Compared with the rigidly connected bistable system, the elastically connected system can makes large-amplitude oscillations easier, which is beneficial to improve the performance of the energy harvester, especially suitable for energy harvesting in a low frequency environment.
3. The spring stiffness has an important effect on the performance of the proposed system. A spring with a small spring stiffness is beneficial for the system to achieve a large-amplitude oscillation over a wider frequency band. However, in practical applications, the spring stiffness affects the position of the equilibrium points of the system, the minimum spring stiffness must be able to maintain the bistable characteristics of the system, which is a problem that must be considered in the design. Otherwise, the elastically connected bistable system will lose its bistable

characteristics and degenerate into a nonlinear monostable system, thus resulting in poor energy harvesting performance.

In addition, experimental investigations will be presented in the future.

Author Contributions: Conceptualization, X.Z. and X.C.; Methodology, X.C.; Software, Y.G.; Validation, F.Z., X.C. and L.C.; Writing—original draft preparation, X.C.; Writing—review and editing, L.C.; Visualization, Y.G.; Supervision, X.C.; Project administration, X.Z.; Funding acquisition, X.Z. All authors have read and agreed to the published version of the manuscript.

Funding: This research was funded by the National Natural Science Foundation of China (Grant No. 51974228 and No. 51834006), the National Green Manufacturing System Integration Project (Grant No. 2017-327), Shaanxi Innovative Talent Plan Project (Grant No. 2018TD-032), and Key R&D project in Shaanxi (Grant No. 2018ZDCXL-GY-06-04).

Institutional Review Board Statement: Not applicable.

Informed Consent Statement: Not applicable.

Data Availability Statement: Not applicable.

Acknowledgments: Authors acknowledge the support from the students Lin Wang, Meng Zuo and Xiao She for their assistance in the modeling and experiments.

Conflicts of Interest: The authors declare no conflict of interest.

References

1. Wang, J.; Geng, L.; Zhou, S.; Zhang, Z.; Lai, Z.; Yurchenko, D. Design, modeling and experiments of broadband tristable galloping piezoelectric energy harvester. *Acta Mech. Sin.* **2020**, *36*, 592–605. [CrossRef]
2. Daqaq, M.F.; Masana, R.; Erturk, A.; Quinn, D.D. On the role of nonlinearities in vibratory energy harvesting: A Critical review and discussion. *Appl. Mech. Rev.* **2014**, *66*, 040801. [CrossRef]
3. Yildirim, T.; Ghayesh, M.H.; Li, W.; Alici, G. A review on performance enhancement techniques for ambient vibration energy harvesters. *Renew. Sustain. Energy Rev.* **2017**, *71*, 435–449. [CrossRef]
4. Karadag, C.V.; Ertarla, S.; Topaloglu, N.; Okyar, F. Optimization of beam profiles for improved piezoelectric energy harvesting efficiency. *Struct. Multidiscip. Optim.* **2021**, *63*, 631–643. [CrossRef]
5. Tao, K.Z.; Chen, Z.S.; Yi, H.P.; Zhang, R.R.; Shen, Q.; Wu, J.; Tang, L.H.; Fan, K.Q.; Fu, J.Q.; Miao, J.M.; et al. Hierarchical honeycomb-structured electret/triboelectric nanogenerator for biomechanical and morphing wing energy harvesting. *Nano-Micro Lett.* **2021**, *13*, 1–16. [CrossRef]
6. Zou, H.-X.; Zhao, L.-C.; Gao, Q.-H.; Zuo, L.; Liu, M.; Tan, T.; Wei, K.-X.; Zhang, W.-M. Mechanical modulations for enhancing energy harvesting: Principles, methods and applications. *Appl. Energy* **2019**, *255*, 113871. [CrossRef]
7. Yang, Z.; Zhou, S.; Zu, J.; Inman, D. High-performance piezoelectric energy harvesters and their applications. *Joule* **2018**, *2*, 642–697. [CrossRef]
8. Zhang, X.; Yang, W.; Zuo, M.; Tan, H.; Fan, H.; Mao, Q.; Wan, X. An Arc-shaped piezoelectric bistable vibration energy harvester: Modeling and experiments. *Sensors* **2018**, *18*, 4472. [CrossRef]
9. Rubes, O.; Brablc, M.; Hadas, Z. Nonlinear vibration energy harvester: Design and oscillating stability analyses. *Mech. Syst. Signal Process.* **2019**, *125*, 170–184. [CrossRef]
10. Erturk, A.; Inman, D.J. Assumed-modes formulation of piezoelectric energy harvesters: Euler-Bernoulli, Rayleigh and Timoshenko models with axial deformations. In Proceedings of the ASME 2010 10th Biennial Conference on Engineering Systems Design and Analysis, Istanbul, Turkey, 12–14 July 2010; pp. 405–414. [CrossRef]
11. Stanton, S.C.; McGehee, C.C.; Mann, B. Nonlinear dynamics for broadband energy harvesting: Investigation of a bistable piezoelectric inertial generator. *Phys. Nonlinear Phenom.* **2010**, *239*, 640–653. [CrossRef]
12. Li, K.; Yang, Z.; Zhou, S. Performance enhancement for a magnetic-coupled bi-stable flutter-based energy harvester. *Smart Mater. Struct.* **2020**, *29*, 085045. [CrossRef]
13. Singh, K.A.; Kumar, R.; Weber, R.J. A broadband bistable piezoelectric energy harvester with nonlinear high-power extraction power electronics. *IEEE Trans.* **2015**, *30*, 6763–6774.
14. Zhou, Z.; Qin, W.; Du, W.; Zhu, P.; Liu, Q. Improving energy harvesting from random excitation by nonlinear flexible bi-stable energy harvester with a variable potential energy function. *Mech. Syst. Signal Process.* **2019**, *115*, 162–172. [CrossRef]
15. Cao, J.; Wang, W.; Zhou, S.; Inman, D.J.; Lin, J. Nonlinear time-varying potential bistable energy harvesting from human motion. *Appl. Phys. Lett.* **2015**, *107*, 143904. [CrossRef]
16. Nguyen, M.S.; Yoon, Y.-J.; Kwon, O.; Kim, P. Lowering the potential barrier of a bistable energy harvester with mechanically rectified motion of an auxiliary magnet oscillator. *Appl. Phys. Lett.* **2017**, *111*, 253905. [CrossRef]

17. Yang, W.; Towfighian, S. Internal resonance and low frequency vibration energy harvesting. *Smart Mater. Struct.* **2017**, *26*, 095008. [CrossRef]
18. Yang, W.; Towfighian, S. Low frequency energy harvesting with a variable potential function under random vibration. *Smart Mater. Struct.* **2018**, *27*, 114004. [CrossRef]
19. Yang, W.; Towfighian, S. A parametric resonator with low threshold excitation for vibration energy harvesting. *J. Sound Vib.* **2019**, *446*, 129–143. [CrossRef]
20. Lan, C.; Qin, W. Enhancing ability of harvesting energy from random vibration by decreasing the potential barrier of bistable harvester. *Mech. Syst. Signal Process.* **2017**, *85*, 71–81. [CrossRef]
21. Shan, G.; Wang, D.F.; Song, J.; Fu, Y.; Yang, X. A spring-assisted adaptive bistable energy harvester for high output in low-excitation. *Microsyst. Technol.* **2018**, *24*, 3579–3588. [CrossRef]
22. Li, X.; Li, Z.; Huang, H.; Wu, Z.; Huang, Z.; Mao, H.; Cao, Y. Broadband spring-connected bi-stable piezoelectric vibration energy harvester with variable potential barrier. *Results Phys.* **2020**, *18*, 103173. [CrossRef]
23. Kim, J.; Dorin, P.; Wang, K.W. Vibration energy harvesting enhancement exploiting magnetically coupled bistable and linear harvesters. *Smart Mater. Struct.* **2020**, *29*, 065006. [CrossRef]
24. Qian, F.; Zhou, S.; Zuo, L. Approximate solutions and their stability of a broadband piezoelectric energy harvester with a tunable potential function. *Commun. Nonlinear Sci. Numer. Simul.* **2020**, *80*, 104984. [CrossRef]
25. Ooi, B.L.; Gilbert, J.M.; Aziz, A.R.A. Analytical and finite-element study of optimal strain distribution in various beam shapes for energy harvesting applications. *Acta Mech. Sin.* **2016**, *32*, 670–683. [CrossRef]
26. Xie, X.; Wang, Z.; Liu, D.; Du, G.; Zhang, J. An experimental study on a novel cylinder harvester made of L-shaped piezoelectric coupled beams with a high efficiency. *Energy* **2020**, *212*, 118752. [CrossRef]
27. Wang, B.; Luo, X.; Liu, Y.; Yang, Z. Thickness-variable composite beams for vibration energy harvesting. *Compos. Struct.* **2020**, *244*, 112232. [CrossRef]
28. Yang, Z.; Wang, Y.Q.; Zuo, L.; Zu, J. Introducing arc-shaped piezoelectric elements into energy harvesters. *Energy Convers. Manag.* **2017**, *148*, 260–266. [CrossRef]
29. Yung, K.W.; Landecker, P.B.; Villani, D.D. An analytic solution for the force between two magnetic dipoles. *Magn. Electr. Sep.* **1998**, *9*, 39–52. [CrossRef]
30. Hartog, D.J. *Mechanical Vibrations*; McGraw-Hill: New York, NY, USA, 1956.
31. Erturk, A.; Inman, D.J. *Piezolectric Energy Harvesting*; John Wiley & Sons: Hoboken, NJ, USA, 2011.

micromachines

MDPI

Article

Design and Performance Evaluation of a Single-Phase Driven Ultrasonic Motor Using Bending-Bending Vibrations

Dongmei Xu [1,*], Wenzhong Yang [1], Xuhui Zhang [1] and Simiao Yu [2,*]

1 Shaanxi Key Laboratory of Mine Electromechanical Equipment Intelligent Monitoring, Xi'an University of Science and Technology, Xi'an 710054, China; ywz18503483941@163.com (W.Y.); zhangxh@xust.edu.cn (X.Z.)
2 School of Mechanical and Electrical Engineering, Xi'an University of Architecture and Technology, Xi'an 710055, China
* Correspondence: dongmeixu@xust.edu.cn (D.X.); simiaoyu@xauat.edu.cn (S.Y.)

Abstract: An ultrasonic motor as a kind of smart material drive actuator has potential in robots, aerocraft, medical operations, etc. The size of the ultrasonic motor and complex circuit limits the further application of ultrasonic motors. In this paper, a single-phase driven ultrasonic motor using Bending-Bending vibrations is proposed, which has advantages in structure miniaturization and circuit simplification. Hybrid bending vibration modes were used, which were excited by only single-phase voltage. The working principle based on an oblique line trajectory is illustrated. The working bending vibration modes and resonance frequencies of the bending vibration modes were calculated by the finite element method to verify the feasibility of the proposed ultrasonic motor. Additionally, the output performance was evaluated by experiment. This paper provides a single-phase driven ultrasonic motor using Bending-Bending vibrations, which has advantages in structure miniaturization and circuit simplification.

Keywords: ultrasonic motor; single-phase driven; bending vibration

Citation: Xu, D.; Yang, W.; Zhang, X.; Yu, S. Design and Performance Evaluation of a Single-Phase Driven Ultrasonic Motor Using Bending-Bending Vibrations. *Micromachines* **2021**, *12*, 853. https://doi.org/10.3390/mi12080853

Academic Editors: Kai Tao and Yunjia Li

Received: 30 June 2021
Accepted: 19 July 2021
Published: 21 July 2021

1. Introduction

The ultrasonic motor is a kind of special motor based on the inverse piezoelectric effect, which has the merits of no electromagnetic interference, no need for lubrication, fast response, and high positioning accuracy [1–3]. Thus, the ultrasonic motor has been used in fields such as camera lens drives, robots, optical fiber connections, biomedical engineering, etc. [4–6].

From the viewpoint of the phase number of excitation power supply, the ultrasonic motor can be divided into single-phase driven ones [7–9], two-phase driven ones [10,11], and multi-phase driven ones [12,13]. As the phase shift of each phase of excitation voltages should be adjustable, the power supplies of the two-phase-driven and multi-phase-driven ultrasonic motors are relatively complex and large. Tian et al. proposed a single-phase-driven piezoelectric actuator, which worked with an eight-shaped trajectory, and the piezo rings were clamped between the flange bolt and the horn [8]. A single-phase-driven piezoelectric actuator using the longitudinal bending coupling mode was proposed by Liu et al.; when one signal voltage with the frequency of the first longitudinal and third bending resonance frequency was applied to the motor and the boundary was unsymmetrical, oblique elliptical movement was generated to push the mover [14]. However, the consistency demand of the frequency of this longitudinal bending hybrid mode is relatively high. Flueckiger et al. proposed a single-phase ultrasonic motor, in which the longitudinal vibration mode was converted to the particular deformation of the resonator [15]. However, to obtain the forward motion, a signal frequency of 84 kHz was utilized, and to achieve the backward motion, a signal frequency of 69 kHz was used; thus, the output mechanical characteristics of the bi-directional motions are not consistent.

Based on the structure of the metal base and piezoelectric element, ultrasonic motors can be divided into two types: bonded type [16–18] and sandwich type [19–21]. The sandwich type ultrasonic motor has the advantages of large output force and high velocity. The sandwich type Langevin transducer in the literature [19] had an output mechanical force of 92 N and a no-load velocity of 0.47 m/s. A frog-shaped sandwich type piezoelectric actuator in the literature [20] achieved a maximum speed and a thrust of 287 mm/s and 11.8 N, respectively. However, because of the existence of stud structure, the structure of the sandwich type ultrasonic motor is relatively large. Therefore, in some specific situations, the use of the sandwich type ultrasonic motor is restricted.

In view of the above situations, a novel single-phase-driven bonded type ultrasonic motor is proposed in this study, which is beneficial to the miniaturization of motor size and drive circuit. Bending-Bending vibrations are utilized to form the desired oblique line driving trajectory. Additionally, there is no need for frequency degeneracy of Bending-Bending vibrations in this study. Section 2 introduces the structure and working principle of the single-phase-driven ultrasonic motor. Finite element analysis of the single-phase-driven ultrasonic motor is illustrated in Section 3. Output performance of this single-phase-driven bonded type ultrasonic motor is evaluated in Section 4. Finally, the conclusion is provided.

2. Structure and Working Principle of the Single-Phase-Driven Ultrasonic Motor

The structure of the proposed single-phase-driven ultrasonic motor is shown in Figure 1a, which is composed of one aluminum alloy base and four pieces of PZT ceramic. The integrated base has three functioning parts, which are the base, horn, and driving foot. The horn is designed to magnify the vibration amplitude. In order to demonstrate the two orthogonal bending vibration modes of the ultrasonic motor, the polarization direction of four pieces of PZT ceramic is illustrated in Figure 1b. The bonded type of the proposed ultrasonic motor makes it suitable for miniaturization.

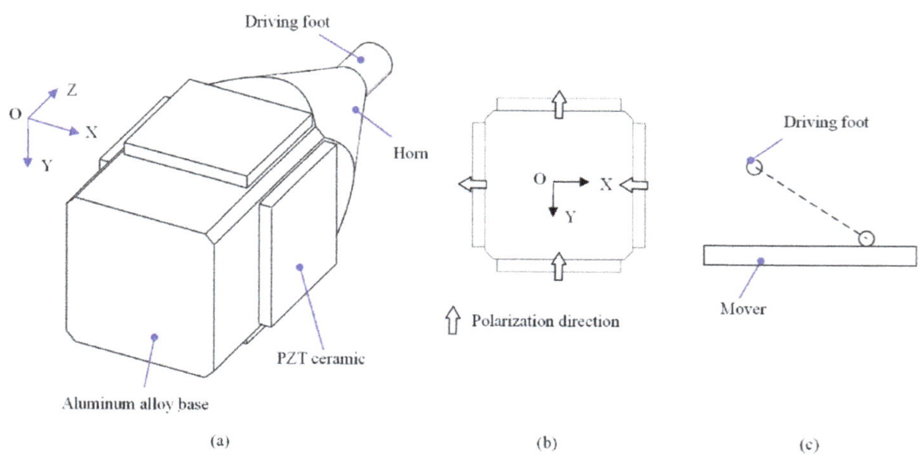

Figure 1. Structure and working principle of the proposed single-phase-driven ultrasonic motor: (**a**) structure, (**b**) polarization directions, (**c**) schematic diagram of driving trajectory.

In addition, in a traditional case of a hybrid of two orthogonal bending vibration modes, two sinusoidal excitation voltages with a phase shift of 90 degrees are used to form an elliptical driving trajectory [22,23]. In this study, two orthogonal bending vibration modes with a 0-degree phase shift are utilized; thus, displacements in the OX and OY directions will be generated simultaneously; then the oblique line driving trajectory is formed, as shown in Figure 1c. Under the proposed principle, only single-phase excitation voltage is needed, which is beneficial to reduce the power cost, simplify the circuit, and miniaturize the whole ultrasonic motor.

3. Finite Element Analysis of the Single-Phase-Driven Ultrasonic Motor

The finite element method is used to calculate the vibration modes and to obtain the resonance frequencies of the Bending-Bending vibration modes. The finite element method (FEM) model of the proposed ultrasonic motor was built in ANSYS, as shown in Figure 2. The numbers of nodes and elements of the FEM model are 58657 and 40698, respectively. The element type of the FEM model is SOLID227. The properties of the aluminum alloy and the PZT ceramics are listed in Table 1.

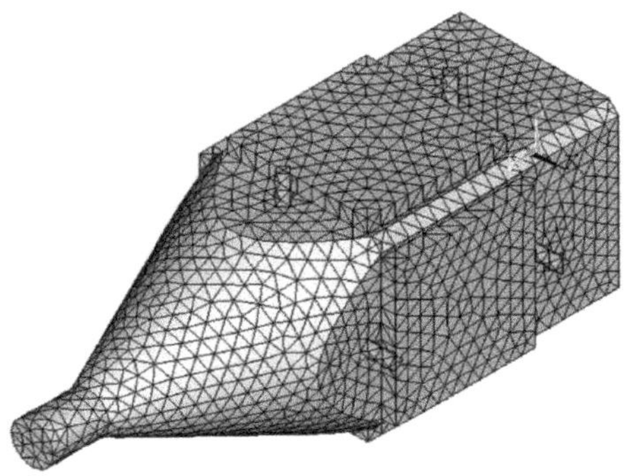

Figure 2. The FEM model of the proposed ultrasonic motor built in ANSYS.

Table 1. The properties of the aluminum alloy and the PZT ceramics.

PZT41	Aluminum Alloy
Piezoelectric matrix $d = \begin{bmatrix} 0 & 0 & 0 & 0 & 5 & 0 \\ 0 & 0 & 0 & 5 & 0 & 0 \\ -1.6 & -1.6 & 3.3 & 0 & 0 & 0 \end{bmatrix} \times 10^{-10}$ C/N	Density ρ = 2810 kg/m^3
Stiffness matrix $c^E = \begin{bmatrix} 15 & 8.4 & 6.8 & 0 & 0 & 0 \\ 8.4 & 15 & 6.8 & 0 & 0 & 0 \\ 6.8 & 6.8 & 12.9 & 0 & 0 & 0 \\ 0 & 0 & 0 & 3.3 & 0 & 0 \\ 0 & 0 & 0 & 0 & 2.8 & 0 \\ 0 & 0 & 0 & 0 & 0 & 2.8 \end{bmatrix} \times 10^{10}$ N/m^2	Poisson's ratio μ = 0.33
Dielectric matrix $\varepsilon^T = \begin{bmatrix} 8.1 & 0 & 0 \\ 0 & 8.1 & 0 & 0 \\ 0 & 0 & 6.7 \end{bmatrix} \times 10^{-9}$ F/m	Modulus of elasticity E = 4.72 GPa

The optimized size of the proposed ultrasonic motor is achieved by parameter sensitivity analysis to ensure that the working frequency of the motor is greater than 20 kHz and the amplitude of bending vibration is greater than 1 μm. The total length of the aluminum alloy base is 36 mm, the height of the cross section of the base is 12 mm, the diameter of the driving foot is 3 mm, and the length of the horn is 15 mm. The size of the PZT ceramic is $10 \times 10 \times 1$ mm^3, and the position of the PZT ceramic is shown in Figure 3. The detailed dimensions of the single-phase-driven ultrasonic motor using Bending-Bending vibrations are shown in Figure 3.

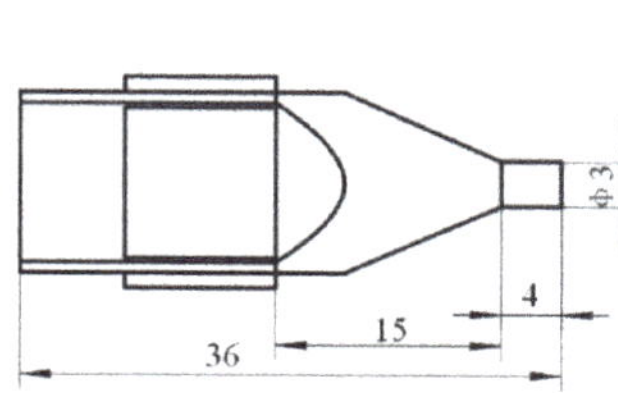

Figure 3. The dimensions of the proposed ultrasonic motor (unit: mm).

The calculated bending vibration modes in OX and OY directions are shown in Figure 4, the resonance frequencies are 41,023 Hz and 41,107 Hz, respectively, and the main reason for the frequency deviation is the unsymmetrical mesh of the model. The calculated driving trajectory is shown in Figure 5, which is an oblique line as proposed in Section 2. In addition, a clamping device was designed, which is shown in Section 4; in the FEM model, a displacement constraint was applied to the ultrasonic motor by cylinders to simulate the constraint of the clamping device.

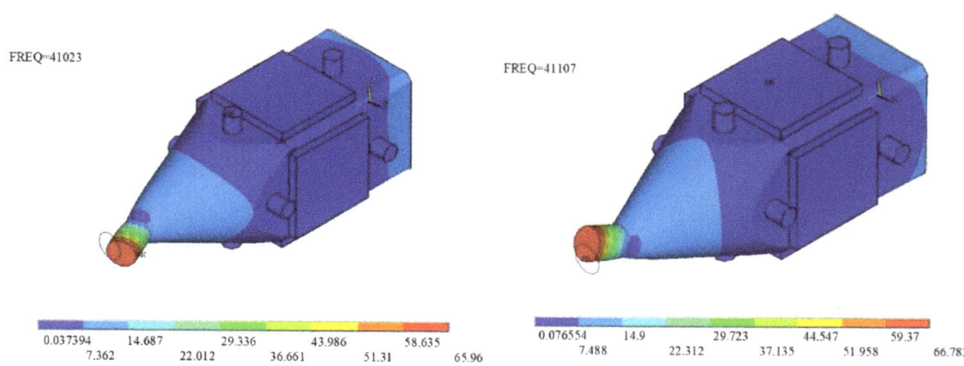

Figure 4. Bending vibration modes in OX and OY directions.

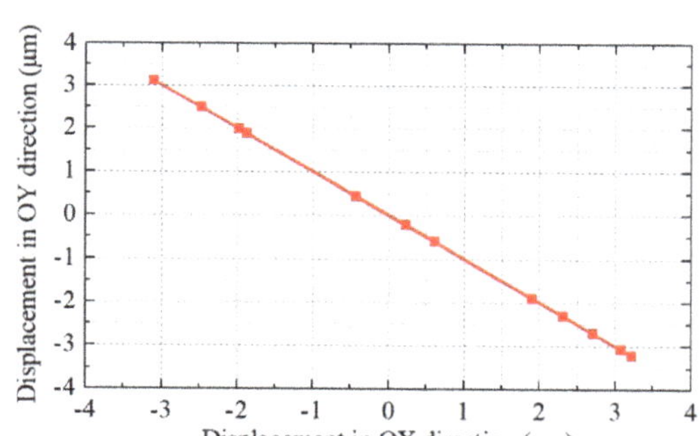

Figure 5. The calculated driving trajectory.

4. Mechanical Characteristics of the Single-Phase-Driven Ultrasonic Motor

In order to evaluate the output performance, a prototype was manufactured. The stator of the ultrasonic motor was composed of one integrated aluminum alloy base and four pieces of PZT ceramic, the dimensions of which were the same as the optimized simulation results. Additionally, four PZT ceramics were pasted on the stator surface with resin glue at the positions shown in Figure 3, and the curing time was 24 h under the action of preload. The impedance characteristics were tested by an impedance analyzer (ZX80A, Zhixin Precision Electronics Co., Ltd., Changzhou, China), as shown in Figure 6. The tested resonance frequency was 41.92 kHz, and the deviation of the simulation resonance frequency and the test one was 855 Hz, which was approximately 2.1% of the simulation resonance frequency. The main reasons for the deviation are the parameters error of the aluminum alloy base and the PZT ceramics, the manufacturing error, and the error caused by the test condition of the impedance analyzer.

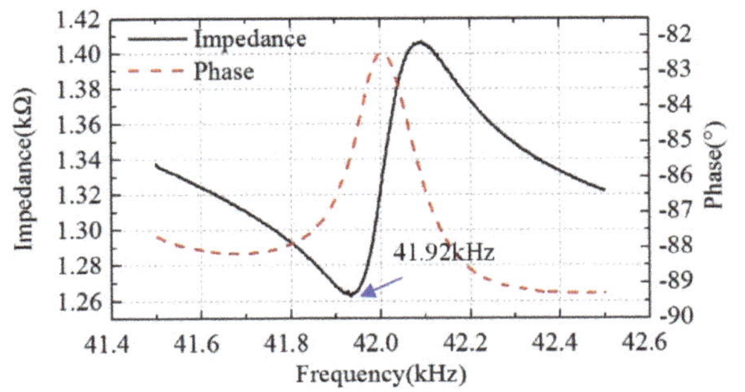

Figure 6. The impedance characteristics.

Then, the output performance was tested under the single-phase excitation voltage. The clamping device and the experimental setup are shown in Figure 7. The prototype was clamped and fixed on the foundation support; the driving foot was pressed on the linear guide rail. The excitation voltage was generated by the signal generator, then amplified by s power amplifier (ATA-4051, Agitek, China); the single-phase excitation voltage, sine signal, was applied to the PZT ceramics of the proposed ultrasonic motor.

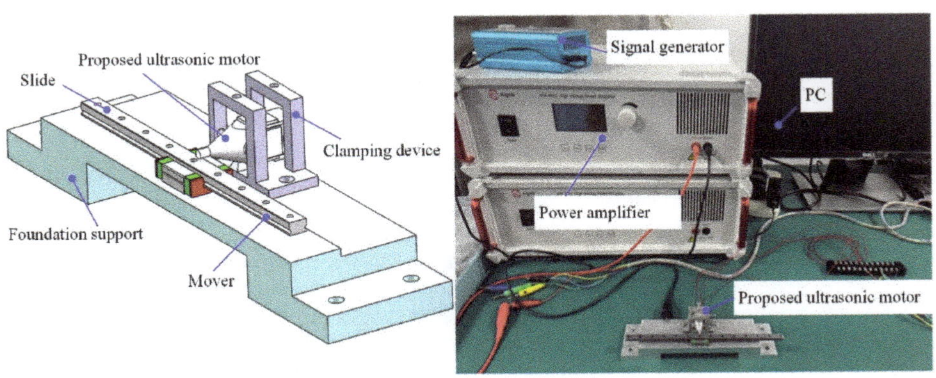

Figure 7. The clamping device and the experimental setup.

The output velocity versus the input excitation voltage frequency is shown in Figure 8. We can see that changing the frequency is another way to change the output velocity. The maximum output velocity of the mover was achieved at a frequency of 42.1 kHz. As the ultrasonic motor works in a resonance state, when the working frequency is far from the resonance frequency, the output velocity decreases rapidly.

The output velocity versus the input excitation voltage amplitude is shown in Figure 9, which indicates that we can change the voltage amplitude to increase the output velocity. With the excitation voltage no more than 120 V, the mover cannot be driven. The maximum velocity was approximately 340 mm/s under an excitation voltage of 300 V and 42.1 kHz.

In addition, the proposed single-phase ultrasonic motor using Bending-Bending vibration modes is feasible, which can also output rotary motion if the linear guide rail is replaced by a ring. The proposed bonded-type single-phase-driven ultrasonic motor not only has the merit of easy miniaturization, but also has a simple and easy miniaturization circuit. This single-phase-driven ultrasonic motor is indeed an impact motor, which has potential to be used in a high-accuracy platform.

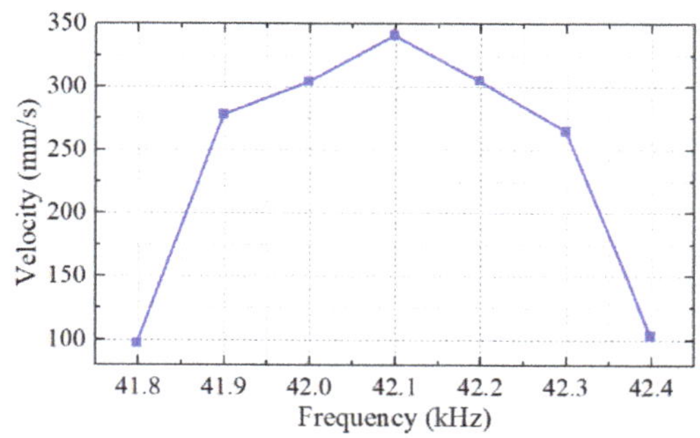

Figure 8. The output velocity versus the input excitation voltage frequency.

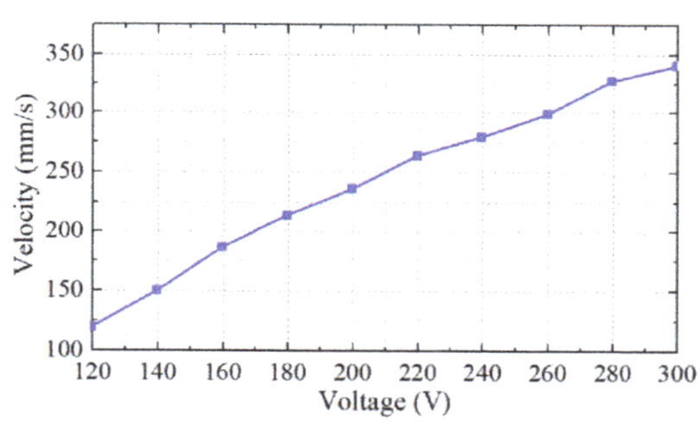

Figure 9. The output velocity versus the input excitation voltage amplitude.

5. Conclusions

A single-phase-driven ultrasonic motor using Bending-Bending vibrations was proposed in this paper. The structure of this ultrasonic motor was composed of a metal base and four pieces of PZT ceramic. Additionally, orthogonal bending vibration modes were excited simultaneously by only single-phase voltage, thus an oblique line driving trajectory was formed to drive the mover. The working principle was verified by the finite element method. Additionally, the impedance characteristics of the ultrasonic motor were tested. The output performance was evaluated by experiment. Additionally, the maximum output velocity under 300 $V_{p\text{-}p}$ was 340 mm/s. The practicability of this proposed single-phase-driven ultrasonic motor was verified. This paper provides a single-phase-driven ultrasonic motor, which has merits in the miniaturization of structures and power circuits. In future work, we will focus on the verification of the linear trajectory and its application in a high-accuracy platform.

Author Contributions: Conceptualization, D.X. and S.Y.; methodology, D.X.; software, W.Y.; validation, W.Y. and D.X.; formal analysis, S.Y.; investigation, W.Y.; resources, D.X.; data curation, W.Y.; writing—original draft preparation, S.Y.; writing—review and editing, W.Y.; visualization, D.X.; supervision, D.X.; project administration, X.Z.; funding acquisition, X.Z. All authors have read and agreed to the published version of the manuscript.

Funding: This work was supported in part by the National Natural Science Foundation of China (No. 52005398), in part by the China Postdoctoral Science Foundation (No. 2019M663776), in part by the Shaanxi Natural Science Basic Research Program (No. 2019JQ-805), in part by the Shaanxi Education Department General Special Scientific Research Plan (No. 20JK0774), and in part by the Shaanxi Key Laboratory of Mine Electromechanical Equipment Intelligent Monitoring for the Open Fund (SKL-MEEIM201916).

Conflicts of Interest: The authors declare no conflict of interest.

References

1. Liu, J.; Liu, Y.; Zhao, L.; Xu, D.; Chen, W.; Deng, J. Design and experiments of a single-foot linear piezoelectric actuator operated in a stepping mode. *IEEE Trans. Ind. Electron.* **2018**, *65*, 8063–8071. [CrossRef]
2. Liu, Y.; Wang, L.; Gu, Z.; Quan, Q.; Deng, J. Development of a two-dimensional linear piezoelectric stepping platform using longitudinalbending hybrid actuators. *IEEE Trans. Ind. Electron.* **2018**, *6*, 3030–3040.
3. Liu, Y.; Yan, J.; Wang, L.; Chen, W. A two-DOF ultrasonic motor using a longitudinal–bending hybrid sandwich transducer. *IEEE Trans. Ind. Electron.* **2018**, *66*, 3041–3050. [CrossRef]
4. Wan, X.; Liu, M.; Zhang, X.; Fan, H.; Mao, Q.; Dong, M.; Wang, X.; Ma, H. Excitation and propagation of longitudinal l (0, 2) mode ultrasonic guided waves for the detection of damages in hexagonal pipes: Numerical and experimental studies. *Shock Vib.* **2021**, *2021*, 6641828.
5. Lok, M.; Helbling, E.F.; Zhang, X.; Wood, R.; Brooks, D.; Wei, G.Y. A low mass power electronics unit to drive piezoelectric actuators for flying microrobots. *IEEE Trans. Power Electron.* **2017**, *33*, 3180–3191. [CrossRef]
6. Tao, K.; Chen, Z.; Yi, H.; Zhang, R.; Shen, Q.; Wu, J.; Tang, L.; Fan, K.; Fu, Y.; Miao, J.; et al. Hierarchical Honeycomb-Structured Electret/Triboelectric Nanogenerator for Biomechanical and Morphing Wing Energy Harvesting. *Nano-Micro Lett.* **2021**, *13*, 123. [CrossRef] [PubMed]
7. Yokoyama, K.; Tamura, H.; Masuda, K.; Takano, T. Single-phase drive ultrasonic linear motor using a linked twin square plate vibrator. *Jpn. J. Appl. Phys.* **2013**, *52*, 07HE03. [CrossRef]
8. Tian, X.; Liu, Y.; Deng, J.; Chen, W. Single-phase drive bending-bending piezoelectric actuator operated under 8-shaped trajectory vibration: Concept, computation and experiment evaluation. *Mech. Syst. Signal Process.* **2020**, *139*, 106637. [CrossRef]
9. Cao, T.; Li, X.; Wen, Z.; Wang, B.; Liu, W.; Yu, D.; Wu, D. A CFRP/PZT laminated piezoelectric motor with high force density. *Smart Mater. Struct.* **2021**, *30*, 075005. [CrossRef]
10. Jiang, Z.; Wang, L.; Jin, J. A novel robotic arm driven by sandwich piezoelectric transducers. *Ultrasonics* **2018**, *84*, 373–381. [CrossRef]
11. Aoyagi, M.; Tomikawa, Y. Ultrasonic rotary motor using longitudinal and bending multimode vibrator with mode coupling caused by external additional asymmetry. *Jpn. J. Appl. Phys.* **1993**, *32*, 4190. [CrossRef]
12. Zhang, X.; Zhang, G.; Nakamura, K.; Ueha, S. A robot finger joint driven by hybrid multi-DOF piezoelectric ultrasonic motor. *Sens. Actuators A Phys.* **2011**, *169*, 206–210. [CrossRef]
13. Takemura, K.; Ohno, Y.; Maeno, T. Design of a plate type multi-DOF ultrasonic motor and its self-oscillation driving circuit. *IEEE/ASME Trans. Mechatron.* **2004**, *9*, 474–480. [CrossRef]
14. Liu, Y.; Shi, S.; Li, C.; Chen, W.; Liu, J. A novel standing wave linear piezoelectric actuator using the longitudinal-bending coupling mode. *Sens. Actuators A Phys.* **2016**, *251*, 119–125. [CrossRef]
15. Flueckiger, M.; Fernandez, J.; Giljum, M.; Perriard, Y. Optimization of a single phase ultrasonic linear motor. *Proc. IEEE Ultrason. Symp.* **2007**, 2327–2330.
16. Wang, L.; Wang, X.; Twiefel, J.; Jin, J.; Xu, Q. Design of a novel surface-bonded type of linear ultrasonic motor with single-phase excitation. In Proceedings of the 2020 IEEE 16th International Conference on Automation Science and Engineering (CASE), Hong Kong, China, 20–21 August 2020; pp. 1078–1083.
17. Shi, Y.; Zhao, C. A new standing-wave-type linear ultrasonic motor based on in-plane modes. *Ultrasonics* **2011**, *51*, 397–404. [CrossRef]
18. Lim, K.J.; Kwon, O.D.; Park, C.H.; Lee, J.S.; Kang, S.H. The design and characteristics of a novel ultrasonic motor. *Mater. Sci. Forum* **2006**, *514*, 1457–1461. [CrossRef]
19. Yun, C.H.; Ishii, T.; Nakamura, K.; Ueha, S.; Akashi, K. A high power ultrasonic linear motor using a longitudinal and bending hybrid bolt-clamped Langevin type transducer. *Jpn. J. Appl. Phys.* **2001**, *40*, 3773. [CrossRef]
20. Zhang, Q.; Chen, W.; Liu, Y.; Liu, J.; Jiang, Q. A frog-shaped linear piezoelectric actuator using first-order longitudinal vibration mode. *IEEE Trans. Ind. Electron.* **2016**, *64*, 2188–2195. [CrossRef]
21. Yamaguchi, D.; Kanda, T.; Suzumori, K. An ultrasonic motor for cryogenic temperature using bolt-clamped Langevin-type transducer. *Sens. Actuators A Phys.* **2012**, *184*, 134–140. [CrossRef]
22. Xu, D.; Liu, Y.; Liu, J.; Chen, W. A bonded type ultrasonic motor using the bending of a crossbeam. *IEEE Access.* **2016**, *4*, 1109–1116. [CrossRef]
23. Liu, Y.; Yang, X.; Chen, W.; Yao, Y. A high power linear ultrasonic motor using a bending bolt-clamped transducer. *Ferroelectrics* **2013**, *445*, 39–50. [CrossRef]

micromachines

Article

Hydraulic Integrated Interconnected Regenerative Suspension: Modeling and Characteristics Analysis

Sijing Guo [1,2], Liang Chen [1,2], Xikai Wang [1,2], Junyi Zou [3] and Sanbao Hu [1,2,*]

1 School of Automotive Engineering, Wuhan University of Technology, Wuhan 430070, China; sijingg@whut.edu.cn (S.G.); 262030@whut.edu.cn (L.C.); wxk704051999@foxmail.com (X.W.)
2 Hubei Key Laboratory of Advanced Technology for Automotive Components, Wuhan University of Technology, Wuhan 430070, China
3 School of Automotive and Traffic Engineering, Wuhan University of Science and Technology, Wuhan 430070, China; zoujunyi@wust.edu.cn
* Correspondence: husanbao@whut.edu.cn

Abstract: A novel suspension system, the hydraulic integrated interconnected regenerative suspension (HIIRS), has been proposed recently. This paper demonstrates the vibration and energy harvesting characteristics of the HIIRS. The HIIRS model is established as a set of coupled, frequency-dependent equations with the hydraulic impedance method. The mechanical–fluid boundary condition in the double-acting cylinders is modelled as an external force on the mechanical system and a moving boundary on the fluid system. By integrating the HIIRS into a half car model, its free and forced vibration analyses are conducted and compared with an equivalent traditional off-road vehicle. Results show that the natural frequency and the damping ratio of the HIIRS-equipped vehicle are within a proper range of a normal off-road vehicle. The root mean square values of the bounce and roll acceleration of the HIIRS system are, respectively, 64.62 and 11.21% lower than that of a traditional suspension. The average energy harvesting power are 186.93, 417.40 and 655.90 W at the speeds of 36, 72 and 108 km/h for an off-road vehicle on a Class-C road. The results indicate that the HIIRS system can significantly enhance the vehicle dynamics and harvest the vibration energy simultaneously.

Keywords: energy harvesting; hydraulic energy-harvesting shock absorber; hydraulic interconnected suspension; vibration characteristics; vehicle dynamics

Citation: Guo, S.; Chen, L.; Wang, X.; Zou, J.; Hu, S. Hydraulic Integrated Interconnected Regenerative Suspension: Modeling and Characteristics Analysis. *Micromachines* **2021**, *12*, 733. https://doi.org/10.3390/mi12070733

Academic Editors: Kai Tao and Yunjia Li

Received: 13 May 2021
Accepted: 18 June 2021
Published: 22 June 2021

1. Introduction

Since the issue of energy depletion was raised, energy harvesting technology has received extensive attention [1–3]. In the automotive industry, extensive research and applications, including braking energy harvesting [4], exhaust thermo-electric recovery [5] and suspension vibration energy recovery [6–8], have also been conducted. Among them, braking energy recovery were gradually applied in electric vehicles. Considerable research has been carried out on energy harvesting suspensions. Zuo et al. [9] proved that the harvestable power in the suspensions of a middle-sized vehicle was 100–400 W, when driven at 60 mph on good and average roads. Zhang et al. [10] harvested 46 W from the suspension system of a passenger car. The research indicated that the energy harvesting suspension is feasible in theory and practice.

Among all traffic accidents, a large number of accidents are due to vehicle rollovers. It is reported that over one-third of 4WD fatalities involved rollover [11]. A percentage of 18.9 of rollover-related fatal crashes happened in the year of 2014 in the United States [12]. Suspension systems play a key role in reducing vehicle roll rates, and superior suspensions can greatly reduce the vehicular rolling propensity. Hydraulic interconnected suspension (HIS), with lower cost and easier structure than semi-active and active suspensions, was verified to be useful in reducing the roll rate [13] and, thus, became a potential alternative to the transmission of conventional suspensions.

To simultaneously enhance the vehicle safety and harvest energy, novel suspensions, which combine energy regeneration and hydraulic interconnection features, have been proposed recently and are of great value to the automotive industry.

1.1. Interconnected Suspensions

The "interconnected" idea of suspension was initially proposed by Hawley in 1927 as "oil pipe interconnected shock absorber" [14]. Since then, various layouts of interconnected suspension and research on the corresponding modeling have been carried out. At present, the main layouts of hydraulic interconnected suspensions are anti-roll [15], anti-pitch [16], anti-bounce and diagonal layout [17], etc. Among them, the most widely studied one is the anti-roll interconnected suspension, whose modeling is highlighted here.

Zhang et al. [18] proposed a frequency domain method to study the hydraulic interconnected suspension system, derived the coupled frequency-related equations and obtained the free vibration solution and frequency response function of a half-vehicle system. Wang et al. [19] did further research. He applied this method to a 7-degree-of-freedom model and studied the effects of several parameters on the roll, pitch and bounce modes of the vehicle. The results were verified by vehicle drop tests. Wang et al. [20] indicated that the HIS system can improve the vehicle stability in both roll and lateral aspects. Ding et al. [16] extended the application of the HIS system to tri-axle trucks, and the modal analysis results indicated that the HIS system can reduce the pitching motion of the sprung mass while maintaining smoothness.

Time-domain analyses on interconnected suspensions have also been conducted. Zhang et al. [21] proposed a new damper for hydraulically interconnected suspension. The AMESim model of hydraulic interconnected suspension was established and simulated. The results showed that the setting time and the overshoot were, respectively, reduced by 42.2 and 14.7%, and the largest roll angle and steady roll were, respectively, decreased by 9.9 and 5.9°. Wang et al. [22] derived the vehicle dynamics model of the hydraulically interconnected inertial device-spring-damper suspension (HIISDS). Two road excitations were used to verify the effectiveness of the suspension. Cao et al. [23] developed a generalized 14 degree-of-freedom nonlinear vehicle model to evaluate the vehicle dynamics of the interconnected suspension. The results showed that the coupled hydro-pneumatic suspension had considerable potential in enhancing the ride comfort and anti-roll/pitch performances.

The existing modeling methods of the mechanical–hydraulic coupled vehicle systems were either in time domain [21–23] or in frequency domain [18–20]. The frequency-domain modeling research mostly utilized the transfer matrix method to evaluate the impedance matrix of the hydraulic subsystem. With the model in the frequency domain, the modal analysis of the system could be performed and the system's own characteristics could be analyzed. The corresponding results disclosed a good agreement with experiments [19].

1.2. Energy Harvesting Shock Absorbers

Researchers noticed the energy-dissipating nature of shock absorbers and began to study energy harvesting shock absorbers [24,25]. The number of publications per year on energy regenerative shock absorber has been exponentially increased over the last decade [26], including piezo-electric [27] and electromagnetic [28,29] energy harvesting techniques. Among all the energy harvesting shock absorbers, the hydraulic energy harvesting shock absorber was put into the application earlier than other types, attributable to its relatively simple structure [30]. Wu et al. [31] established a mathematical model of the hydraulic energy regenerative shock absorber and conducted a series of bench tests. The results showed that the peak recovery power reached 505.52 W, and the recovery efficiency was 14.5%. Samn et al. [32] designed a hydraulic regenerative shock absorber and demonstrated that it cannot only collect part of the wasted energy in the suspension system, but also improve the vehicle ride comfort and road holding. Fang et al. [33] presented a hydraulic electromagnetic energy-regenerative shock absorber and demonstrated that the damping force varied with the magnitude of the load current. Guo et al. [34,35] demonstrated the method and steps for the size matching and parameter setting of the hydraulic components in the hydraulic electromagnetic energy-harvesting shock absorber (HESA) system. Test results indicated that the proper matching of the parameters can make the HESA system work efficiently and improve the energy conversion efficiency. Peng et al. [36] applied the HESA in a commercial vehicle suspension and indicated that the root mean square energy regenerative power were 41.72 and 339.88 W on Class-B and Class-C roads at speeds of 30–70 km/h.

1.3. Energy Harvesting Hydraulic Interconnected Suspension

To combine the merits of the hydraulic interconnected suspension and the energy harvesting shock absorber, various types of the energy harvesting hydraulic interconnected suspension came into individuals' sight, with the aim to simultaneously enhance the vehicle dynamics performance and harvest energy.

Chen et al. [37] integrated the energy harvesting shock absorbers into the hydraulic interconnected suspension system and indicated that the system performed better than the traditional suspension in terms of rolling dynamics and could harvest 421 W energy at 4 Hz and 40 mm (peak) excitation. Guo et al. [38] proposed a hydraulic interconnected suspension system based on hydraulic electromagnetic shock absorbers, which only adopted one set of hydraulic motor-generator system and greatly reduced the cost while improving the energy recovery efficiency. The hydraulic integrated interconnected regenerative suspension (HIIRS) [39], which is studied in this paper, consists of a two-way hydraulic cylinder installed between the wheel and the body, an oil pipe connecting the hydraulic cylinders, a high-pressure accumulator, a low-pressure accumulator, two hydraulic rectifiers and a hydraulic motor-generator unit. Its structure and working principle are shown in Figure 1.

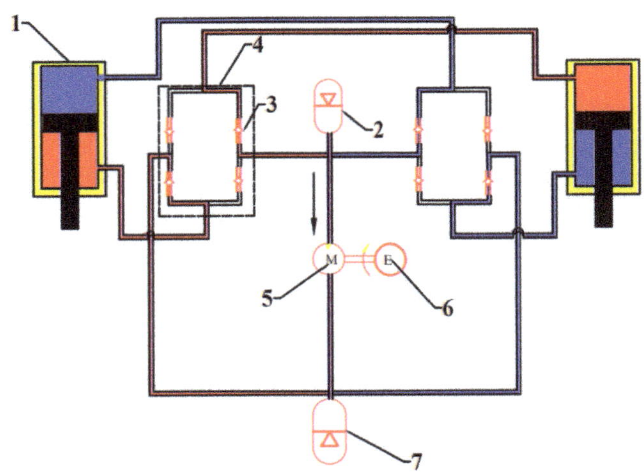

Figure 1. The structure and working principle of the HIIRS. 1 hydraulic cylinder; 2 high-pressure accumulator; 3 check valve; 4 hydraulic rectifier; 5 hydraulic motor; 6 generator; 7 low-pressure accumulator.

The hydraulic rectifier, composed of four check valves, ensures the one-way flow of fluid to drive the hydraulic motor. There is a high-pressure accumulator at the inlet of the hydraulic motor, which can stabilize the fluid flow through the hydraulic motor. This ensures the hydraulic motor to maintain a stable speed and the generator to generate electricity efficiently. The low-pressure accumulator at the outlet can compensate for the variation of the fluid volume in the HIIRS system. The high-pressure fluid flows through the port of the high-pressure accumulator, and thus, when the high-pressure accumulator is working, it provides an extra rigidity to the suspension.

Recent studies on HIIRS mainly focused on design, modeling and experiments. Their results proved that the energy regenerative suspension had certain energy harvesting capability while ensuring comfort. However, the existing research did not model the HIIRS in the frequency domain. The frequency domain modeling method can easily allow for obtaining the evaluation index of vehicle ride comfort and the energy harvesting power at different road surfaces and vehicle speeds. Compared with the time domain analysis method, the frequency domain analysis method has several distinct advantages when applied to HIIRS. (1) The solution is simpler and more convenient; (2) the natural frequency can be calculated, which could guide the future design of the HIIRS; (3) the frequency domain analysis method can demonstrate the responses of the HIIRS under various excitations in an expedient way. In this paper, a half car model coupled with an HIIRS system is developed in the frequency domain. Based on the model, the vibration isolation characteristics and energy harvesting power of HIIRS are studied.

The rest of the paper is organized as follows. Section 2 develops the model of the HIIRS-equipped half vehicle in the frequency domain based on the block modeling and hydraulic impedance method. Free vibration analysis and forced vibration analysis are performed in Section 3. Energy harvesting power is calculated and estimated in Section 4. Finally, Section 5 concludes this paper.

2. Modeling

In this chapter, the idea of modular modeling is adopted. As the HIIRS is introduced into the vehicle, the whole system is divided into two parts: the mechanical system and the hydraulic system. The two parts and their boundary coupling conditions are discussed separately, and finally, the coupled dynamic equation is obtained.

2.1. Mechanical System

Considering the simplicity of modeling, while still accounting for fluid interconnections between the wheel stations and the lumped mass, a four-DOF half-car model, as shown in Figure 2, is used in this investigation.

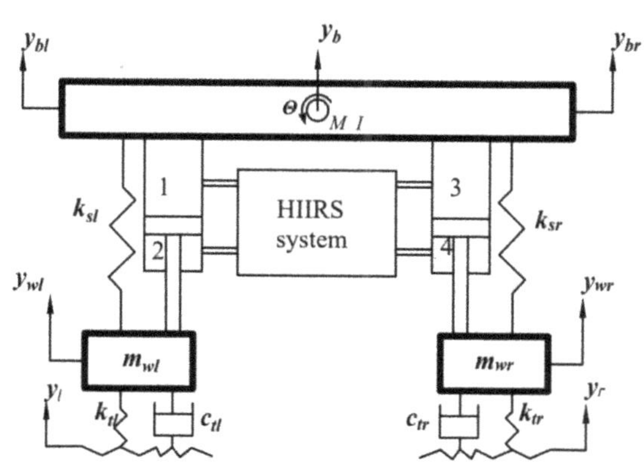

Figure 2. Schematic of a half-car with an HIIRS system.

In this section, the mechanical system is what we concerned about, hence the force of the hydraulic system is regarded as an external force. According to Newton's second law, the kinematics equation of the half-car model is written by

$$M\ddot{y} + C\dot{y} + Ky = f(t) \tag{1}$$

where y is the displacement vector, y = $[y_{wl}, y_{wr}, y_b, \theta]$; $f(t)$ is the resultant force applied to the vehicle, which can be written as $f(t) = D_1 Ap(t) + f_x(t)$. In the equation of $f(t)$, $D_1 Ap(t)$ is the hydraulic cylinder force, and f_x is other external forces vector. We define the pressure vector as $p(t) = [P_1\ P_2\ P_3\ P_4]^T$ and the area matrix as $A = diag(A_1, A_2, A_3, A_4)$. D_1 is a linear transformation matrix $D_1 = \begin{bmatrix} -1 & 1 & 0 & 0 \\ 0 & 0 & -1 & 1 \\ 1 & -1 & 1 & -1 \\ -b_l & b_l & b_r & -b_r \end{bmatrix}$. A_i is the cross-sectional area of the hydraulic cylinder chamber in Figure 1, b_l and b_r are, respectively, the distances from the plane of gravity center of the vehicle to the left and right suspension.

Equation (1) can now be rewritten as

$$M\ddot{y} + C\dot{y} + Ky = D_1 Ap(t) + f_x(t) \tag{2}$$

where $M = \begin{bmatrix} m_l & 0 & 0 & 0 \\ 0 & m_r & 0 & 0 \\ 0 & 0 & M & 0 \\ 0 & 0 & 0 & I \end{bmatrix}, C = \begin{bmatrix} c_{tl} & 0 & 0 & 0 \\ 0 & c_{tr} & 0 & 0 \\ 0 & 0 & 0 & 0 \\ 0 & 0 & 0 & 0 \end{bmatrix}$,

$$K = \begin{bmatrix} k_{sl} + k_{tl} & 0 & -k_{sl} & b_l k_{sl} \\ 0 & k_{sr} + k_{sl} & -k_{sr} & -b_r k_{sr} \\ -k_{sl} & -k_{sr} & k_{sl} + k_{sr} & -b_l k_{sl} + b_r k_{sr} \\ b_l k_{sl} & -b_r k_{sr} & -b_l k_{sl} + b_r k_{sr} & b_l^2 k_{sl} + b_r^2 k_{sr} \end{bmatrix}.$$

2.2. Mechanical–Fluid System Boundary Conditions

In a double-acting hydraulic actuator cylinder, any piston movement will cause liquid to flow into or out of the chamber of the cylinder. Assuming the pressure difference between the upper and lower chambers produces linear leakage at the piston [40], then the linear leakage of the left and right actuators, q_l and q_r, are given by

$$\begin{cases} q_l = \frac{p_{v1} - p_{v2}}{R_l} \\ q_r = \frac{p_{v3} - p_{v4}}{R_r} \end{cases} \tag{3}$$

where R_l and R_r are linearized loss coefficients.

The mechanical–fluid boundary condition, shown in Figure 3, are given by

$$\begin{cases} q_{v1} = q_{A1} - q_l \\ q_{v2} = q_{A2} - q_l \\ q_{v3} = q_{A3} - q_r \\ q_{v4} = q_{A4} - q_r \end{cases} \tag{4}$$

where $q_{Ai}(i = 1\,2\,3\,4)$ is the fluid volume caused by piston motion, which equals the product of the relative speed of the piston v and the piston area A.

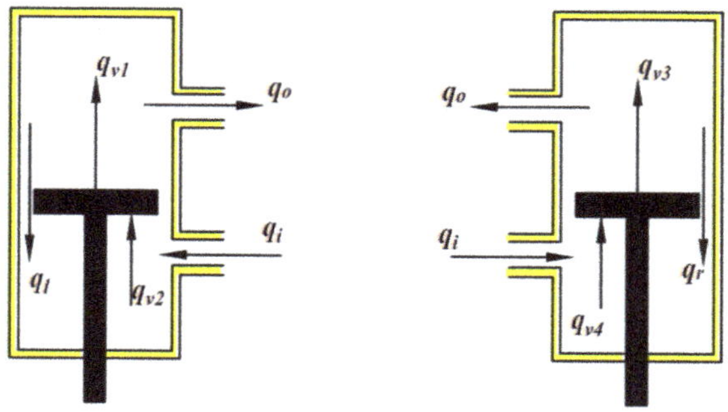

Figure 3. The mechanical–fluid system boundary condition.

For a small roll angle, the relative speed of left and right pistons is

$$\begin{cases} v_l = \dot{y_{wl}} - \dot{y_b} + b_l\dot{\theta} \\ v_r = \dot{y_{wr}} - \dot{y_b} - b_r\dot{\theta} \end{cases} \tag{5}$$

Combining Equations (3)–(5), there are

$$q(t) = AD_2\dot{y} + Rp(t) \tag{6}$$

where $p(t)$ is the flow vector, $p(t) = [P_1\ P_2\ P_3\ P_4]^T$. The matrix D_2 and R are defined as

$$D_2 = \begin{bmatrix} 1 & 0 & -1 & b_l \\ 1 & 0 & -1 & b_l \\ 0 & 1 & -1 & -b_r \\ 0 & 1 & -1 & -b_r \end{bmatrix}, R = \begin{bmatrix} -1/R_l & 1/R_l & 0 & 0 \\ -1/R_l & 1/R_l & 0 & 0 \\ 0 & 0 & -1/R_r & 1/R_r \\ 0 & 0 & -1/R_r & 1/R_r \end{bmatrix}.$$

2.3. Fluid System

In order to solve Equations (2) and (6), which relate to the mechanical system and mechanical–fluid boundary, the fluid system equation in the form $q = f(p)$ must be obtained. $f(p)$ depends on the modelling approach used. For the sake of computational efficiency and analytical advantages, only the linear function between p and q is considered. In particular, the focus of this study is frequency domain modeling. Therefore, the target is to seek the linear relationship between the flow rate $Q(s)$ and the pressure $P(s)$ in the frequency domain.

According to the definition of hydraulic impedance, the relationship between the flow rate and the pressure of a fluid system can be expressed as

$$Q(s) = Z(s)^{-1}P(s) \tag{7}$$

where Z is the impedance matrix.

Equation (6) describes the fluid system of the HIIRS, while Equation (7) describes the mechanical system of the HIIRS. Next, we must combine the two systems, to solve the HIIRS. By comparing the Laplace form of Equations (6) and (7), we can obtain the relationship between the pressure $P(s)$ and the exciting displacement $Y(s)$, as

$$P(s) = sE(s)AD_2Y(s) \tag{8}$$

where $E(s) = \left(Z(s)^{-1} - R\right)^{-1}$. Without considering leakage, $E(s) = Z(s)$.

The Laplace transform form of Equation (2) is the system differential equation in the frequency domain

$$Ms^2Y(s) + sCY(s) + KY(s) = sD_1AE(s)AD_2Y(s) + F_x(s) \tag{9}$$

$$[s^2M + s\overline{C}(s) + K]Y(s) = F_x(s) \tag{10}$$

where $\overline{C}(s) = C - D_1AE(s)AD_2$.

Defining the state vector $x = [y,\ sy]^T$, Equation (10) can be rewritten as

$$sX(s) = \hat{A}(s)X(s) + \hat{B}U(s) \tag{11}$$

where $\hat{A}(s) = \begin{bmatrix} 0 & I \\ -M^{-1}K & -M^{-1}\overline{C}(s) \end{bmatrix}$, $\hat{B} = \begin{bmatrix} 0 \\ M^{-1} \end{bmatrix}$, $U(s) = F_x(S)$.

It should be noted that it is not easy to determine the impedance correlation matrix $\hat{A}(s)$, which depends on the fluid circuit and the arrangement of the various components. The circuit layout in this study adopts anti-oppositional interconnection, as shown in Figure 4.

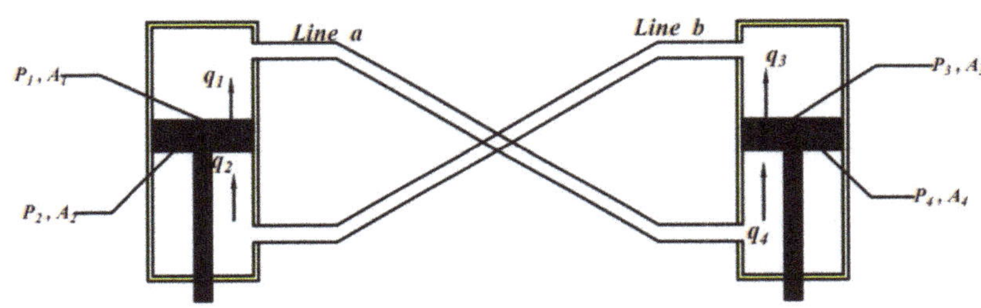

Figure 4. Schematic of a general anti-oppositional half-car HIIRS arrangement.

Using the original boundary flow definitions, q_i, the nodal state vectors for the mechanical–fluid interfaces in this arrangement are related by

$$\begin{bmatrix} P_4 \\ Q_4 \end{bmatrix} = \begin{bmatrix} T_{11}^a & T_{12}^a \\ T_{21}^a & T_{22}^a \end{bmatrix} \begin{bmatrix} P_1 \\ Q_1 \end{bmatrix} \begin{bmatrix} P_2 \\ Q_2 \end{bmatrix} = \begin{bmatrix} T_{11}^b & T_{12}^b \\ T_{21}^b & T_{22}^b \end{bmatrix} \begin{bmatrix} P_3 \\ Q_3 \end{bmatrix} \tag{12}$$

Combining Equation (12) with Equation (7), $Z(s)^{-1}$ can be written as

$$Z(s)^{-1} = \begin{bmatrix} -\frac{T_{22}^a}{T_{21}^a} & 0 & 0 & \frac{1}{T_{21}^a} \\ 0 & \frac{T_{11}^b}{T_{21}^b} & T_{12}^b - \frac{T_{11}^b T_{22}^b}{T_{21}^b} & 0 \\ 0 & \frac{1}{T_{21}^b} & \frac{-T_{22}^b}{T_{21}^b} & 0 \\ T_{12}^a - \frac{T_{11}^a T_{22}^a}{T_{21}^a} & 0 & 0 & \frac{T_{11}^a}{T_{21}^a} \end{bmatrix} \tag{13}$$

By substituting Equation (13) into Equation (12), it yields the complete system equations.

For Equation (13), the values of the elements in the matrix T^a and T^b can be obtained by the hydraulic impedance method and the transfer matrix method. In this method, the state vector of the adjacent state node is related to the transfer matrix T. If the state vector is defined as fluid pressure P and flow rate Q, then

$$\begin{bmatrix} P \\ Q \end{bmatrix}_o = \begin{bmatrix} T_{11} & T_{12} \\ T_{21} & T_{22} \end{bmatrix} \begin{bmatrix} P \\ Q \end{bmatrix}_i \tag{14}$$

where the subscript o represents the fluid output node, and i represents the fluid input node. As long as the previous output node is regarded as the next input node, according to the fluid flow direction in the circuit, the start node and the end node in the circuit can be connected.

As the layout of the hydraulic circuit is shown in Figure 5, the transfer matrix T^a can be expressed as

$$T^a(s) = \mathrm{T}_{X11\to X12} T_{X10\to X11} T_{X9\to X10} T_{X8\to X9} \mathrm{T}_{X7\to X8} T_{X6\to X7} \mathrm{T}_{X5\to X6} T_{X4\to X5} \mathrm{T}_{X3\to X4} \mathrm{T}_{X2\to X3} \mathrm{T}_{X1\to X2} \quad (15)$$

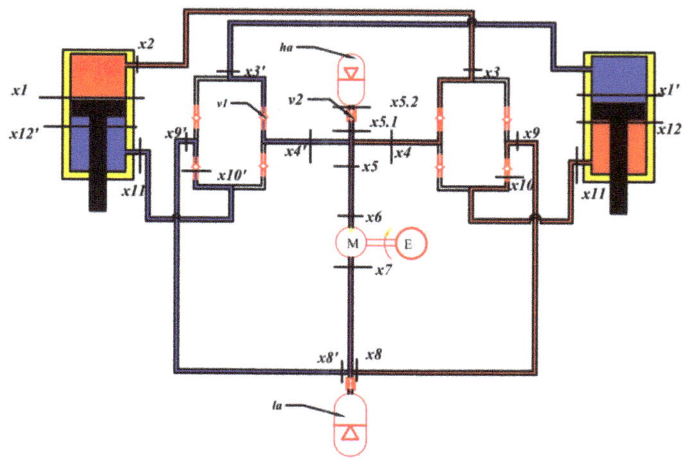

Figure 5. Schematic of a typical half-car with the HIIRS system. v1—check valves; la—low-pressure accumulator; ha—high-pressure accumulator; v2—accumulator valve; x, x′—state nodes.

The transfer matrix of each segment in the circuit depends on the characteristics of the component, which are modeled as follows.

The two-dimensional viscous compressible flow model is applied to model the pipeline, according to [41–43]. The basic fluid equations of the model can be summarized into the equation of state, continuity equation and momentum equation:

Equation (16) of state:

$$\frac{dp}{d\rho} = a_0^2 \quad (16)$$

Continuity Equation (17):

$$\frac{\partial p}{\partial t} + \overline{\rho}\left(\frac{\partial v_x}{\partial x} + \frac{\partial v_r}{\partial r}\frac{v_r}{r}\right) = 0 \quad (17)$$

Momentum Equation (18):

$$\overline{\rho}\frac{\partial v_x}{\partial t} = \frac{-\partial p}{\partial x} + \overline{\mu}\left(\frac{\partial^2 v_x}{\partial r^2} + \frac{1}{r}\frac{\partial v_x}{\partial r}\right) \quad (18)$$

where v_x and v_r are the axial and radial velocity of fluid; $a_0 = \sqrt{\overline{\beta}/\overline{\rho}}$ is the fluid sonic velocity; $\overline{\beta}$, $\overline{\mu}$ and $\overline{\rho}$ are the mean value of fluid bulk modulus, viscosity and density. The field transfer matrix of the pipelines is

$$T_p = \begin{bmatrix} \cosh\Gamma(s) & -Z_C(s)\sinh\Gamma(s) \\ -\frac{\sinh\Gamma(s)}{Z_C(s)} & \cosh\Gamma(s) \end{bmatrix} \quad (19)$$

where $Z_C(s) = \frac{a\overline{\rho}}{A}\left(l - \frac{2J_1\left(ir\sqrt{s/\overline{v}}\right)}{ir\sqrt{s/\overline{v}}J_0\left(ir\sqrt{s/\overline{v}}\right)}\right)^{-1/2}$, $\Gamma(s) = \frac{ls}{a}\left(l - \frac{2J_1\left(ir\sqrt{s/\overline{v}}\right)}{ir\sqrt{s/\overline{v}}J_0\left(ir\sqrt{s/\overline{v}}\right)}\right)^{-1/2}$, and l is the length of the line element, A is the pipeline internal cross-sectional area, r is the pipeline internal radius, $\overline{\rho}$ is the mean fluid density, $\overline{v}$ is the mean fluid kinematic viscosity, and J_0 and J_1 are Bessel functions of the first kind with orders zero and one, respectively. Considering the compressibility of fluid and pipeline, the propagation velocity of transmission line is

$$\mathrm{a} = \sqrt{\frac{a_0^2}{1 + \left(2ra_0^2\overline{\rho}/t_pE\right)}} \tag{20}$$

where t_p is the thickness of the pipe wall, and E is the Young's modulus of the pipe material.

The strict model of the flow dynamics through the damping valves usually involves complicated geometrical parameters. The simplified model used here assumes that the damping valve has a negligible fluid volume. The transfer matrix of the damping valves is then expressed as

$$\Omega_V = \begin{bmatrix} 1 & -Z_v \\ 0 & 1 \end{bmatrix} \tag{21}$$

where $Z_v = R_v$, and R_v is the constant linear pressure loss coefficient.

For the model of the accumulator, the following assumptions have been made. The compressibility of the liquid is far lower than that of the gas in the accumulator; the elasticity of the diaphragm is neglected; there is no heat exchange between the gas in the accumulator and the outside world. The axle moves quickly relative to the small amplitude of the car body, resulting in a rapid reduction or expansion in the volume of the accumulator, and satisfies the gas adiabatic balance equation. At this time, the accumulator can be regarded as a linear system [16], and the impedance of the accumulator is

$$Z_\mathrm{A} = -\frac{\gamma\overline{P}^2}{sp_pv_p} \tag{22}$$

The three-way junctions and the accumulator are connected, and hence, the transfer matrix of the three-way junctions and the accumulator are combined here. According to impedance definition and accumulator impedance

$$P_{X5.2} = Q_{X5.2}Z_\mathrm{A} \tag{23}$$

Set the forward flow direction to be outside the accumulator

$$P_{X5.1} = P_{X5.2} - Z_{V2}Q_{X5.2} \ and \ Q_{X5.2} = Q_{X5.1} \tag{24}$$

Combining Equations (23) and (24), we obtain

$$\frac{P_{X5.1}}{Q_{X5.1}} = Z_A - Z_{V2} = Z_{X5.1} \tag{25}$$

Now, applying the fluid continuity equation at the tee-junction yields

$$Q_{X5} = Q_{X4} + \frac{P_{X5.1}}{Z_{X5.1}} \tag{26}$$

Ignoring the pressure difference between the three nodes, the transfer matrix can be written as

$$\Omega_J = \begin{bmatrix} 1 & 0 \\ \frac{1}{Z_A - Z_{V2}} & 1 \end{bmatrix} \tag{27}$$

The flow rate at the inlet of the hydraulic motor is defined as Q_M. One part of Q_M, which drives the rotation of the hydraulic motor is $\eta_V Q_M$, where η_v is the volumetric efficiency of the hydraulic motor. The other part of the flow is the leakage flow ΔQ_M from the high-pressure cavity to the low-pressure cavity. The flow at the outlet of the hydraulic motor is $\eta_V Q_M$.

When the oil flows through the motor, the relationship between the flow rate Q_m of the hydraulic motor and the speed n_m (rev/s) of the hydraulic motor is

$$n_m = \frac{Q_M \eta_v}{Q_m} \tag{28}$$

Assuming that the equivalent moment of inertia of the motor-generator coupling system is J_m, the speed of the motor is n_m, $n_m = \frac{\omega}{2\pi}$, and ω is the angular velocity. Then, there is a torque balance

$$T_{m2} = T_e + J_m \dot{\omega} \tag{29}$$

With reference to the principle of hydraulic transmission, the input torque T_{m1} and output torque T_{m2} of the hydraulic motor can be calculated according to the following Equation (30)

$$\begin{cases} T_{m1} = \frac{\Delta P_m Q_m}{2\pi} \\ T_{m2} = T_{m1} \eta_m \end{cases} \tag{30}$$

where Δp_m is the pressure drop, η_m is the total efficiency of the hydraulic motor, $\eta_m = \eta_v \eta_p$ is the product of the volumetric efficiency and pressure utilization efficiency.

Combining Equations (26)–(30), we obtain

$$\frac{\Delta P_m Q_m}{2\pi} \eta_m = T_e + 2\pi J_m \dot{n_m} \tag{31}$$

The electromagnetic induction torque T_e of the generator is determined by the armature current I and the torque constant k_t of the motor

$$T_e = k_t I \tag{32}$$

The winding armature current I is related to the design of the energy-regenerative circuit, and the equation can be obtained from Kirchhoff's voltage law, as

$$U_{emf} = k_e \omega \tag{33}$$

Combining Equations (31)–(33) to obtain

$$\Delta p_m = \frac{4\pi^2 k_e k_t \eta_v}{(R_1 + R_2) Q_m^2 \eta_m} Q_M + \frac{4 J_m \pi^2 \eta_v}{Q_m^2 \eta_m} \dot{Q_M} \tag{34}$$

Laplace transform of the above Equation (34) yields

$$\Delta P_m = \frac{4\pi^2 k_e k_t \eta_v}{(R_e + R_{in}) Q_m^2 \eta_m} Q_M + \frac{4 J_m \pi^2 \eta_v}{Q_m^2 \eta_m} s Q_M \tag{35}$$

where R_e is the external resistance, R_{in} is the circuit resistance, k_e is the motor speed constant, Q_M is the motor inlet flow, Q_m is the motor displacement, η_v is the volumetric efficiency, and J_m is the motor-generator rotational inertia.

Therefore, the impedance matrix of the motor-generator unit is

$$Z_M = \frac{\Delta P_M}{Q_M} = \frac{4\pi^2 k_e k_t \eta_v}{(R_e + R_{in}) Q_m^2 \eta_m} + \frac{4 J_m \pi^2 \eta_v}{Q_m^2 \eta_m} s. \tag{36}$$

The transfer matrix of the combined model of hydraulic motor and generator can be written as

$$\begin{bmatrix} P \\ Q \end{bmatrix}_{out} = \begin{bmatrix} 1 & -Z_M \\ 0 & \eta_v \end{bmatrix} \begin{bmatrix} P \\ Q \end{bmatrix}_{in} \tag{37}$$

Now, the hydraulic impedance and transfer matrix of the pipeline, check valve, accumulator and motor-generator are obtained. Substituting Equations (19), (21), (27) and (36) into Equation (15), T^a and T^b can be clear. By substituting the relative elements of T^a and T^b into Equation (13), the matrix $Z^{-1}(s)$ is obtained with definite elements. As a result, the governing equation of the HIIRS system is determined. Then, Equations (10) and (11) can be applied to solve the vibration problem of the HIIRS-equipped half-vehicle.

3. Vibration Analysis

After the above work, the system Equation (11) can be used to analyze the vibration characteristics of the half-car model equipped with HIIRS. The values of each parameter in this study are shown in Table 1.

Table 1. Nomenclature.

Symbol	Values	Units	Description
M	1400	kg	Sprung mass
I	625	kgm^2	Sprung mass moment of inertia about roll axis
m_{wl}, m_{wr}	60	kg	Unsprung mass
b_l, b_r	0.9	m	Distance from gravity center to suspension strut
k_{sl}, k_{sr}	30	kNm^{-1}	Mechanical suspension spring stiffness
k_{tl}, k_{tr}	300	kNm^{-1}	Tire spring stiffness
ρ	870	kgm^{-3}	Density
μ	0.05	Nsm^{-1}	Viscosity
β_{oil}	1400	MPa	Bulk modulus
l	1	m	Length of pipe
d_p	16×10^{-3}	m	Pipeline diameter
V_{ph}	4×10^{-4}	m^3	Pre-charge gas volume of ha
P_{ph}	0.8	MPa	Pre-charge pressure of ha
V_{pl}	4×10^{-4}	m^3	Pre-charge gas volume of la
P_{pl}	0.1	MPa	Pre-charge pressure of la
A_i, A_j	2, 1.6 ($\times 10^{-3}$)	m^2	Upper and lower piston areas (i = 1, 3; j = 2, 4)
s_j	0.15	m	Stroke length
R_{v1}	5×10^8	kgs^{-1}m^{-4}	Linear loss coefficient for check valve
R_{v2}	3×10^8	kgs^{-1}m^{-4}	Linear loss coefficient for accumulator valves
Q_m	10	cc/rev	Hydraulic motor displacement
J_m	0.0005	kgm^2	Motor-generator rotational inertia
k_t	0.25	NM/A	Torque constant
k_e	0.25	V/(rad/s)	Speed constant
R_e	10	Ω	Load Resistance
R_{in}	0.6	Ω	Motor internal resistance

3.1. Free Vibration Analysis

When the external input is zero input, the half-vehicle system vibrates freely, and the corresponding mathematical description is the homogeneous form of Equation (11) as

$$\mathrm{sX(s)} = \hat{A}(s)X(s) \tag{38}$$

The solution of the free vibration of the system can be obtained by solving

$$\det\left(\hat{A}(s) - \mathrm{sI}\right) = 0 \tag{39}$$

To solve Equation (38), we must solve the eigenvalue of $\hat{A}(s)$. Several elements of $\hat{A}(s)$ are functions of s, unless the frequency is known, $\hat{A}(s)$ cannot be completely determined, and therefore, the solution of Equation (38) cannot be obtained by conventional methods. The method used here transforms the process of finding the root of the characteristic equation into the process of finding the local optimal solution.

Assuming that the minimum value of the characteristic equation is the objective function, that is, min $J(s) = |\hat{A}(s) - SI|$ is the objective function, the fminsearch function in MATLAB is used to find the local minimum of the objective function J(s). The specific method is mainly divided into two processes. In the first process, the relatively rough root finding is performed. The value of the objective function is calculated after initializing the Laplacian operator. If the local extremum is not found, the value of the Laplacian operator is reinitialized, and the above process is performed again. If the local extremum is found, then the following second process is performed. The second process is a test process, and the Laplacian found in the first process is substituted into the characteristic matrix to obtain a fixed eigen matrix. The eigenvalue of the fixed value matrix is then solved, and the value is compared with the local extremum found in the first process. If the two values are not equal, skip the second process and return to the first process to find the local eigenvalue again. If the two values are the same, the root finding process ends.

With the determined eigenvalues λ_i and eigenvectors α_i from Equation (38), the natural frequency and damping ratio for each mode are given by

$$f_{n_i} = \frac{\sqrt{(real(\lambda_i))^2 + (imag(\lambda_i))^2}}{2\pi} \text{ and } \xi_i = \frac{|(real(\lambda_i))|}{\sqrt{(real(\lambda_i))^2 + (imag(\lambda_i))^2}} \quad (40)$$

Based on the analysis of the above complex modal vibration theory, as long as the eigenvalues of the characteristic matrix in Equation (37) and the corresponding eigenvectors are solved, the natural frequency, damping ratio and main vibration mode of the system can be obtained. In order to find the roots conveniently, the three-dimensional graph is obtained, whose horizontal and vertical coordinates are the real and imaginary parts of the Laplacian operator, and the vertical coordinates are the objective function to initially determine the number of roots and the range of the roots according to the relevant parameters of the system. The three-dimensional image of the half-vehicle model obtained is shown in Figure 6.

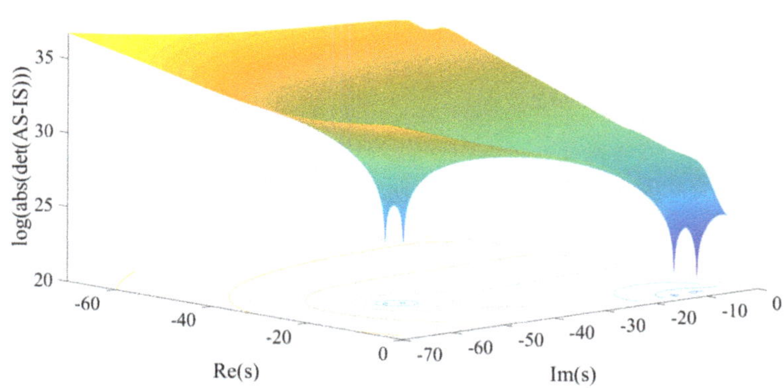

Figure 6. Three-dimensional plot showing the four approximate roots of the characteristic equation of the HIIRS-equipped vehicle.

It is not difficult to see from Figure 6 that the system eigen matrix has four eigenvalues. The system eigenvalues obtained by the local optimization method after initially determining the range of the eigenvalues are shown in Table 2.

Table 2. The approximate eigenvalue solutions to the system matrix of the HIIRS-equipped vehicle.

Eigenvalues	First	Second	Third	Fourth
Real part	−3.7850	−3.6440	−24.4428	−25.7550
Imaginary part	−8.5587	−13.1938	−46.8497	−49.1905

The relatively rough roots obtained in the first process need to be tested in the test process. Substituting the eigenvalues in Table 2 into matrix $\hat{A}(s)$, the system matrix can be completely determined. The traditional methods of finding eigenvalues and eigenvectors can be applied. If the approximate root in process one is equal to the result obtained by the traditional method, it can be considered that the approximate root is the eigenvalue of matrix $\hat{A}(s)$. The natural frequency, damping ratio and mode shape corresponding to the eigenvalues that passed the inspection are shown in the Table 3.

Table 3. Vibration modes of the HIIRS-equipped vehicle.

Mode	Bounce	Roll	Wheel Hop 1 (Synchronous)	Wheel Hop 2 (Oppositional)
Frequency (Hz) and damping ratio	$f_n = 1.49$ $\xi_n = 0.40$	$f_n = 2.18$ $\xi_n = 0.27$	$f_n = 8.41,$ $\xi_n = 0.46$	$f_n = 8.84$ $\xi_n = 0.46$
State variable (s)	$-3.79 \pm 8.56i$	$-3.6 \pm 13.19i$	$-24.44 \pm 46.85i$	$-25.76 \pm 49.19i$
Left wheel (y_{wl})	0.210.23i	−0.29 + 0.18i	1	1
Right wheel (y_{wr})	0.210.23i	0.290.18i	1	−1
Centre of gravity (y_v)	1	0	−0.02 + 0.08i	0
Roll angle (θ)	0	1	0	0.060.14i

The results show that the half-vehicle roll model has four main vibration modes. The first-order vibration mode is dominated by the vertical vibration of the vehicle body and the corresponding natural frequency is 1.49 Hz and the damping ratio is 0.40. Usually, this natural frequency of off-road vehicles is 1.3~2 Hz, and the damping ratio is 0.2~0.4 [44]. The second-order mode is dominated by the body roll vibration. The third-order vibration is dominated by wheel vibration, and the two wheels move in the same direction. The fourth-order vibration mode is dominated by wheel vibration, and the wheels on the left and right sides do reverse vibration. At this time, the body does not vibrate vertically, but has a slight roll vibration.

3.2. *Forced Vibration Analysis*

3.2.1. Frequency Response Matrix

When analyzing the vibration of a vehicle model, the general work to be done is to find the vibration transfer function, which can characterize the amplitude and phase of the vehicle model under different frequency excitations. The vibration transfer matrix of the HIIRS-equipped half-car model is solved as follows.

When the road roughness is used as input, Equation (10) can be obtained as

$$[s^2M + s\overline{C}(s) + K]Y(s) = F_x(s) \tag{41}$$

where $F_x(s) = \vec{F}(s)\vec{\xi}(s)$ is the force exerted by the road on the tire, $\vec{\xi}(s) = [\xi_l, \xi_r, 0, 0]^T$ is the road excitation. $\vec{F}$ is a 4×4 matrix whose elements are zero except for the first two diagonals, which are $\vec{F_{11}}(s) = k_{tl} + sc_{tl}$ and $\vec{F_{22}}(s) = k_{tr} + sc_{tr}$.

Equation (41) can be written as

$$B(s)Y(s) = \overline{F}(s)\overline{\xi}(s) \tag{42}$$

Now, the frequency response matrix of the half-car system can be defined as

$$H_y(s) = \frac{Y(s)}{\bar{\xi}(s)} = B^{-1}(s)\overline{F}(s) \tag{43}$$

With $s = j\omega$, the frequency response matrix describes the system displacement response to any excitations. Therefore, as long as the excitation frequency is known, the HIIRS system equation can be completely determined, and the vibration analysis can be carried out just like other linear vehicle models.

3.2.2. The Response to Random Road Excitation

Frequency domain analysis, similar to time domain analysis, is an important method for studying vehicle system vibration. In the frequency domain analysis map, the independent variable is the frequency and the dependent variable is the amplitude of the signal to be analyzed. In Section 2, the model of the mechanical hydraulic coupling system is established. Now, the input spectrum of the road is necessary to obtain the response of the HIIRS system to the excitation of the road.

Road input is the source of the vehicle vibration system. Whether the impact of the vehicle on the road can be obtained, obtaining accurate road information is the key. In the frequency domain, the power spectral density function is generally used to describe the vibration of the random vibration system. The road power spectrum density function is mainly used in vehicle dynamic response, optimal control of suspension, calculation of road load, etc. The vibration response of the vehicle can be evaluated through the road roughness power spectrum and the dynamic characteristics of the vehicle system. If the frequency response of the suspension is waiting to be solved on the basis of the above transfer function, a model of the road input is also needed. Since the car is excited by the unevenness of the road through the tire contacting the ground, it can be known from the random vibration theory that its vibration response is a smooth random vibration. The research object of this paper is a half-car model. The dynamic response characteristics can be obtained by determining the road input spectrum of the left and right wheels.

Suppose that the spatial frequency n represents the road self-power spectral density, $S_q(n)$ and $n = \frac{1}{\lambda}$ holds. When the vehicle is driving on the road at speed u, it has

$$f = un \tag{44}$$

where f is the time frequency. When the vehicle speed does not change, the time-domain frequency bandwidth Δf has the following relationship with the corresponding spatial domain frequency bandwidth Δn

$$\Delta f = u \cdot \Delta n \tag{45}$$

The power spectral density $S_q(n)$ at the frequency in the spatial domain can be expressed as

$$S_q(n) = \lim_{\Delta n \to 0} \frac{\sigma^2_{q\sim\Delta n}}{\Delta n} \tag{46}$$

where $\sigma^2_{q\sim\Delta n}$ is the energy of the road power spectrum in the frequency domain bandwidth Δn. When the vehicle speed is constant, the harmonic components of the road roughness displacement contained in the time band Δf corresponding to the spatial frequency band Δn are the same, so the road power spectral density in the time domain is

$$S_q(f) = \lim_{\Delta f \to 0} \frac{\sigma^2_{q\sim\Delta n}}{\Delta f} \tag{47}$$

By Equations (44)–(47), $S_q(f)$ can be written as

$$S_q(f) = \frac{1}{u} S_q(n) \tag{48}$$

Under normal circumstances, $S_q(n)$ can be directly calculated. The selection of n in the equation is related to the speed and frequency of the driving vehicle. If the condition that 5 m/s < u< 50 m/s and 0.5 Hz < f < 50 Hz are satisfied at the same time, there is $0.01m^{-1} < n < 10m^{-1}$.

It can be seen from the experiment that $S_q(n)$ can be written as

$$S_q(n) = cn^{-2w} \tag{49}$$

where w is a coefficient ranging from 1–1.25 and is generally taken as 1. The value of c is related to the road surface level, shown in Table 4.

Table 4. The value of c for various road surface level.

Type of Pavement	The Value Range of c
Class A	$8 \times 10^{-8} \sim 32 \times 10^{-8}$
Class B	$32 \times 10^{-8} \sim 128 \times 10^{-8}$
Class C	$128 \times 10^{-8} \sim 512 \times 10^{-8}$

When the frequency index is taken as 2, the road power spectral density $G_q(f)$ satisfies

$$G_q(f) = G_q(n_0)n_0^2\frac{u}{f^2} \tag{50}$$

where the reference spatial frequency n_0 is taken as 0.1. According to Equations (49) and (50), we can obtain

$$S_q(f) = c\frac{u}{f^2} \tag{51}$$

Considering the actual coherence of the left and right wheels, the road input spectral density matrix of the half-vehicle model is

$$\mathrm{S} = \begin{bmatrix} S_D & S_X \\ S_X & S_D \end{bmatrix} \tag{52}$$

where S_D is the self-power spectral density of the road excitation

$$S_D = \frac{1}{u}S_q(n) = \frac{1}{u}\mathrm{c}n^{-2w} \tag{53}$$

The road cross power spectral density S_X is

$$S_X = \frac{1}{u}\left[2\mathrm{c}\left(\frac{\pi L}{n}\right)^w/\Gamma(\mathrm{w})\right]J_w(2\pi Ln) \tag{54}$$

where L is the left and right wheel track, J_w is the second-class modified Bessel function of order w, and $\Gamma(\mathrm{w})$ is the gamma function.

When the left and right wheels are excited by the road roughness transfer function, the relationship between the response spectrum and the input spectrum in the frequency domain is as

$$S_i(f) = [H_{i1}^* \ H_{i2}^*]S\begin{bmatrix} H_{i1} \\ H_{i2} \end{bmatrix} \tag{55}$$

where * represents the conjugate complex number, $S_i(f)$ represents the power spectrum of the i-th output, H_{i1} and H_{i2}, respectively, represent the transfer function of the i-th output to the first and second inputs.

When $c = 64 \times 10^{-8}$ (Class-B road) and the vehicle speed is 36, 72 and 108 Km/h, the power spectrum density of bounce acceleration and the power spectrum density of roll acceleration of the HIIRS system are shown, respectively, in Figure 7.

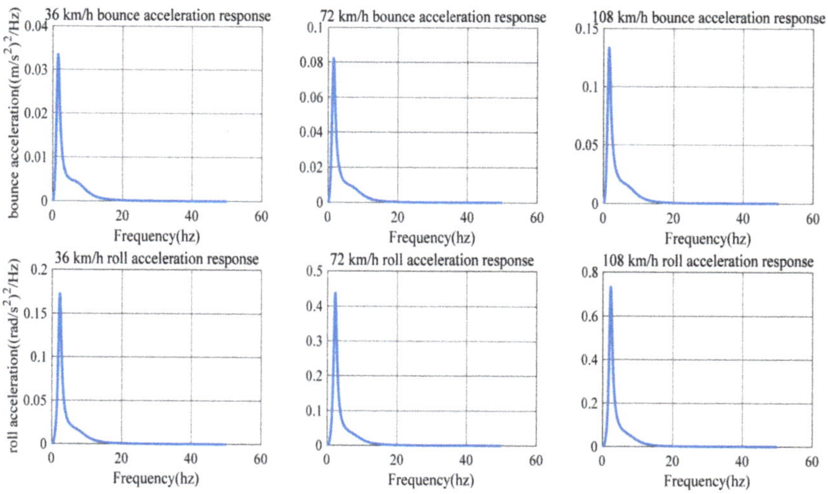

Figure 7. The power spectrum density of the acceleration of an HIIRS-equipped vehicle when driven at 36, 72 and 108 km/h on Class-B road.

When $c = 256 \times 10^{-8}$ (C-level road) and the vehicle speed is 36, 72 and 108 km/h, the power spectrum density of bounce acceleration and the power spectrum density of roll acceleration of the HIIRS system are shown, respectively, in Figure 8.

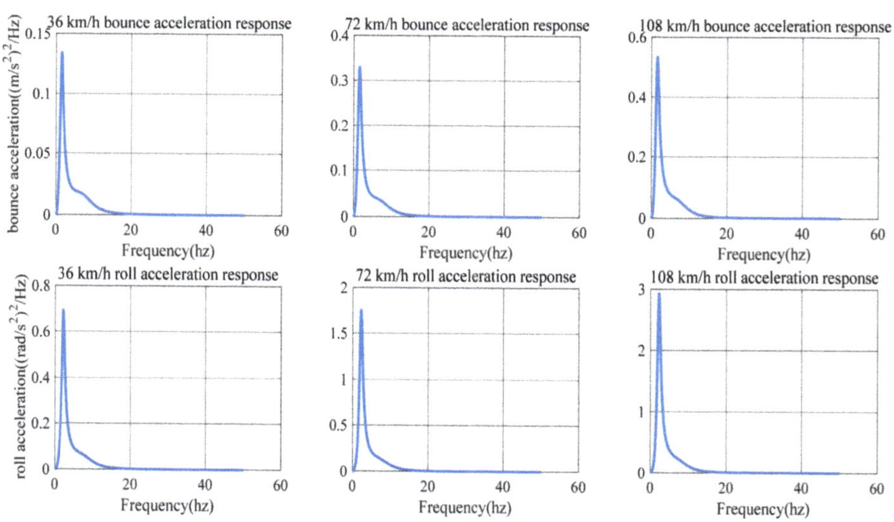

Figure 8. The power spectrum density of the bounce/roll acceleration of an HIIRS-equipped vehicle when driven at 36, 72 and 108 km/h on Class-C road.

Figures 7 and 8 show the power spectrum density of the bounced acceleration and roll acceleration of the vehicle body. The natural frequency and root mean square (RMS) of the acceleration response under Class-B and Class-C roads are in Table 5. It indicates that the natural frequency of bounce is around 1.5 Hz, while the roll frequency is around 2.19 Hz. In addition, an increase in speed on the same road will lead to an increase in the acceleration power spectrum density. At the same speed, an improvement in road conditions will reduce the acceleration power spectrum.

Table 5. The natural frequency and RMS acceleration response of the HIIRS-equipped vehicle.

Working Conditions		Class-B Road			Class-C Road		
		36 km/h	72 km/h	108 km/h	36 km/h	72 km/h	108 km/h
Natural frequency	Bounce (Hz)	1.48	1.50	1.51	1.48	1.50	1.51
	Roll (Hz)	2.19	2.19	2.19	2.19	2.19	2.19
RMS of acceleration	Bounce (m/s^2)	0.0049	0.0120	0.0196	0.0109	0.0480	0.0393
	Roll (rad/s^2)	0.0246	0.0614	0.1029	0.0985	0.2457	0.2059

No matter what kind of road surface excitation, the HIIRS system has the same modal law: no matter the speed is low, medium or high, the natural frequency of the roll mode always maintains at around 2.19 Hz, and the natural frequency of the bounce mode varies when the vehicle speed increases. The natural frequency is 1.48 Hz at low speed, 1.50 Hz at medium speed and 1.51 Hz at high speed. This indicates that when the vehicle speed increases, the HIIRS system can provide greater rigidity. In terms of amplitude, the RMS bounce acceleration at medium speed increased by 143.66% compared with that at low speed, while the RMS bounce acceleration at high speed increased by 63.54% compared with that at medium speed. The RMS roll acceleration at medium speed increased by 149.59% compared with that at low speed, while the RMS roll acceleration at high speed increased by 67.63% compared with that at medium speed. From this point of view, it can be seen that as the vehicle speed increases, the RMS acceleration also increases, but the amount of increase at high speed is less than that at low speed. To a certain extent, it indicates that the anti-bounce and anti-roll capabilities of the HIIRS system are more greatly enhanced at high speeds.

Figure 9 indicates that the acceleration power spectral density curve of the HIIRS system has the same trend with the traditional suspension. The difference is that the peak value of the HIIRS system is much lower than that of the traditional suspension. Additionally, the improvement is more noticeable in the vertical acceleration than the roll acceleration.

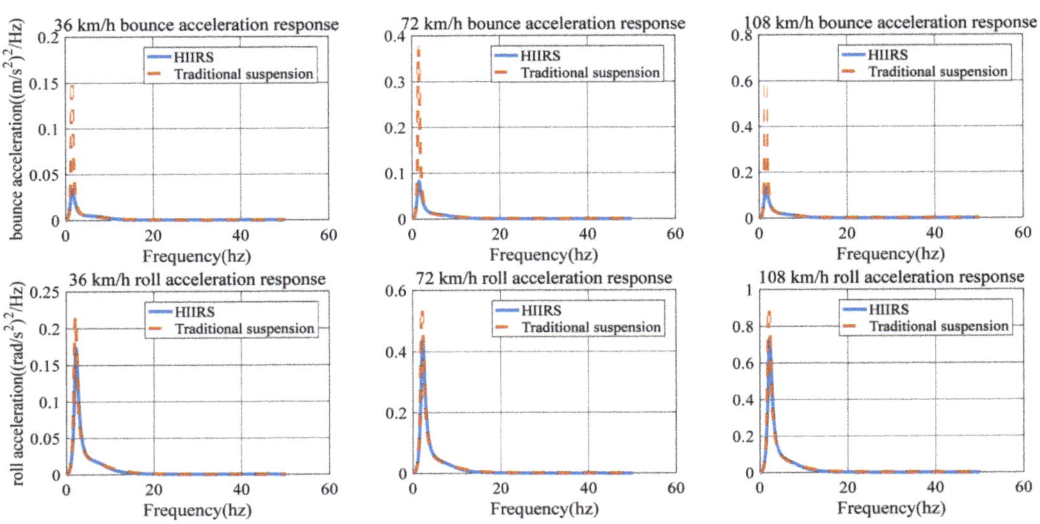

Figure 9. Comparison of bounce/roll acceleration between the HIIRS-equipped vehicle and the traditional vehicle when driven at 36, 72 and 108 km/h on Class-B road.

Figure 9 compared the power spectrum density of bounce/roll acceleration between the HIIRS-equipped vehicle and the traditional vehicle, when the vehicles are driven on a Class-B road ($c = 64 \times 10^{-8}$). In the traditional vehicle, the stiffness and damping are, respectively, 80000 and 4400 Ns/m. The corresponding results are also summarized in Table 6.

Table 6. The comparison of the natural frequency and RMS bounce/roll acceleration between the HIIRS system and traditional suspension system.

Suspension and Speed		Natural Frequency of Bounce (Hz)	Natural Frequency of Roll (Hz)	RMS Bounce Acceleration (m/s^2)	RMS Roll Acceleration (rad/s^2)
HIIRS	36 km/h	1.48	2.19	0.0049	0.0246
	72 km/h	1.50	2.19	0.0120	0.0614
	108 km/h	1.51	2.19	0.0196	0.1029
Traditional suspension	36 km/h	1.48	2.04	0.0139	0.0276
	72 km/h	1.48	2.04	0.0342	0.0690
	108 km/h	1.48	2.04	0.0554	0.1150

From the natural frequency aspect, the roll natural frequency of the HIIRS system and traditional suspension system are maintained at 2.19 and 2.04 Hz, respectively, while the bounce natural frequency of HIIRS ranges from 1.48 to 1.51 HZ and that of traditional suspension is about 1.48 HZ. This indicates that when the road surface and speed are the same, the HIIRS system can provide greater stiffness, especially when the vehicle rolls, since the roll natural frequency of the HIIRS system is higher than that of the traditional suspension.

From the amplitude aspect, Figure 9 shows that the response of the HIIRS is lower than the traditional suspension in the whole frequency range, and the HIIRS can greatly reduce the peak value of the responses. Table 6 shows the RMS of bounce and roll acceleration of the HIIRS system can, respectively, reach 64.91 and 12.38% lower than the traditional suspension when the vehicle is driven at 72 km/h on Class-B road. That is, the HIIRS system provides superior ride comfort and handling stability to traditional suspension systems.

4. Energy Harvesting Power of the HIIRS

Section 3 demonstrated the vibration characteristics of the HIIRS, and this section focuses on the energy harvesting characteristics.

In the energy harvesting circuit, the relationship between the various physical quantities can be written as

$$\begin{cases} P = I^2 R_e \\ I = \frac{U_{emf}}{R_e + R_{in}} \\ U_{emf} = k_e \omega \\ \omega = 2\pi \frac{Q_M}{Q_m} \eta_v \end{cases} \tag{56}$$

where P is the energy harvesting power, R_e is the external resistance, I is the current, U_{emf} is the induced electromotive force, R_{in} is the circuit resistance, k_e is the speed constant of the generator, Q_M is the motor inlet flow, Q_m is the motor displacement, η_v is the volumetric efficiency.

In Equation (56), only Q_M is the unknown quantity, and thus, the goal is to solve Q_M.

According to Figure 4 and Equation (15), the state quantity (pressure, flow) of the motor inlet is

$$T_{X6} = T_{X5 \to X6} T_{X4 \to X5} T_{X3 \to X4} T_{X2 \cdot X3} T_{X1 \cdot X2} T_{X1} \tag{57}$$

State quantity at node 1 is $T_{X1} = \begin{bmatrix} P_1 \\ Q_1 \end{bmatrix}$, P_1 and Q_1 can be calculated according to Equations (7) and (8).

According to Equation (56), the transfer function HX_{61} between the motor inlet flow rate and T_{X1} can be known. Equations (7) and (8) provide the transfer function HX_{1Y}

between T_{X1} and the displacement vector Y. Therefore, the transfer function between the motor inlet flow and the displacement vector Y in Equation (1) is

$$H_{Q_M Y} = HX_{61} \cdot HX_{1Y} \tag{58}$$

Then, the power spectral density of the flow rate at the motor inlet, G_{Q_M}, can be calculated as

$$G_{Q_M} = H_{Q_M Y}^2 G_Y \tag{59}$$

where G_Y is the power spectral density of Y.

The power spectral density has the following relationship with the amplitude

$$G_{Q_M} = A_{Q_M}^2 / f_s \tag{60}$$

where A_{Q_M} represents the amplitude of A_{Q_M}, and f_s represents the frequency bandwidth.

In Section 3, we obtained the power spectral density of bounce and roll acceleration under random road input and the frequency response matrix of the displacement vector Y. If the method of solving acceleration power spectral density is extended to displacement, with the known power spectral density of the road displacement vector Y, the power spectral density of the flow rate at the motor inlet under various random road input can be obtained. Then, the corresponding amplitude can be solved with Equation (60), and the time domain flow rate at the motor inlet can be solved by performing inverse Fourier transform. The time domain flow rate is then substituted into Equation (56), and the energy harvesting power can be determined.

The energy regenerative power of a half-car with HIIRS system on a Class-C road at the speed of 36, 72 and 108 km/h is shown in Figure 10. The average value of energy regenerative power at this road surface excitation is shown in Table 7. It shows that the energy harvesting power can reach 655.90 W for an off-road vehicle when it is driven on a Class-C road at 108 km/h.

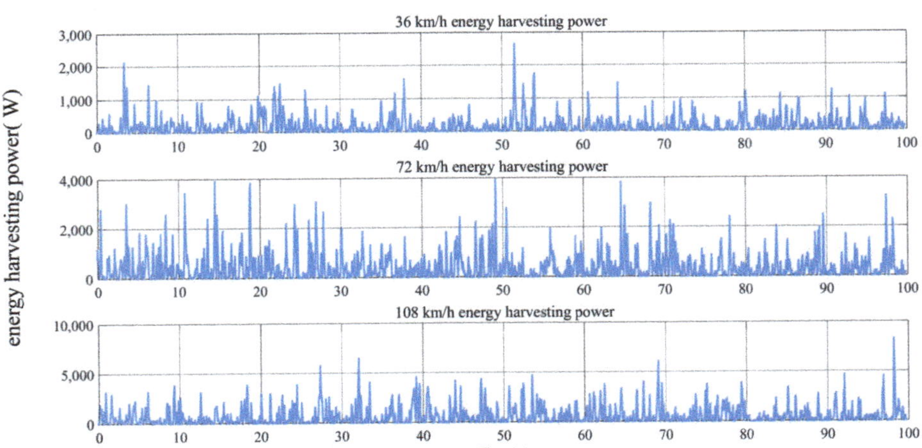

Figure 10. The energy harvesting power of the HIIRS-equipped half-vehicle when driven at various speeds on Class-C road.

Table 7. Energy harvesting power.

Vehicle Speed	36 km/h	72 km/h	108 km/h
Energy harvesting power (W)	186.93	417.39	655.90

5. Conclusions

This paper studied the vibration isolation and energy harvesting characteristics of a novel hydraulic integrated interconnected regenerative suspension (HIIRS). The model in the frequency domain was established. Both free and forced vibration analysis were carried out and compared with a traditional suspension. The comparison showed that the RMS bounce and roll acceleration of the HIIRS system was, respectively, 64.91 and 12.38% lower than the traditional suspension when the vehicle was driven at 72 km/h on a Class-B road. With the frequency domain model of the HIIRS, an approach for calculating the energy harvesting power was also presented. The calculated energy harvesting power was 186.93, 416.40 and 656.90 W, when the vehicle speed was 36, 72 and 108 km/h. In summary, the HIIRS system can significantly enhance the vehicle ride comfort and handling stability while harvesting vibration energy to achieve an energy-saving purpose.

Author Contributions: Conceptualization, S.G.; Funding acquisition, S.G.; Investigation, L.C.; Methodology, S.G. and J.Z.; Software, L.C.; Supervision, S.G. and S.H.; Validation, X.W. and S.H.; Writing—original draft, S.G. and L.C.; Writing—review & editing, S.G., J.Z. and S.H. All authors have read and agreed to the published version of the manuscript.

Funding: This work was funded by National Natural Science Foundation of China (51905394) and Hubei Natural Science Foundation of China (2019CFB202).

Acknowledgments: The authors are grateful to the support from Hubei Key Laboratory of Advanced Technology for Automotive Components, Hubei Collaborative Innovation Centre for Automotive Components Technology.

Conflicts of Interest: The authors declare no conflict of interest. The funders had no role in the design of the study; in the collection, analyses or interpretation of data; in the writing of the manuscript or in the decision to publish the results.

References

1. Zhou, S.; Zuo, L. Nonlinear dynamic analysis of asymmetric tristable energy harvesters for enhanced energy harvesting. *Commun. Nonlinear Sci. Numer. Simul.* **2018**, *61*, 271–284. [CrossRef]
2. Huang, D.; Zhou, S.; Yang, Z. Resonance Mechanism of Nonlinear Vibrational Multistable Energy Harvesters under Narrow-Band Stochastic Parametric Excitations. *Complexity* **2019**, *2019*, 1050143. [CrossRef]
3. Tao, K.; Chen, Z.; Yi, H.; Zhang, R.; Shen, Q.; Wu, J.; Tang, L.; Fan, K.; Fu, Y.; Miao, J.; et al. Hierarchical Honeycomb-Structured Electret/Triboelectric Nanogenerator for Biomechanical and Morphing Wing Energy Harvesting. *Nano Micro Lett.* **2021**, *13*, 123. [CrossRef]
4. Mikołaj, B.; Marcin, P. Multiaspect measurement analysis of breaking energy recovery. *Energy Convers. Manag.* **2016**, *127*, 35–42.
5. Liu, X.; Deng, Y.; Li, Z.; Su, C. Performance analysis of a waste heat recovery thermoelectric generation system for automotive application. *Energy Convers. Manag.* **2016**, *90*, 121–127. [CrossRef]
6. Jia, Y.; Li, S.; Shi, Y. An Analytical and Numerical Study of Magnetic Spring Suspension with Energy Recovery Capabilities. *Energies* **2018**, *11*, 3126. [CrossRef]
7. Jolly, M.R.; Margolis, D.L. Regenerative Systems for Vibration Control. *J. Vib. Acoust.* **1997**, *119*, 208–215. [CrossRef]
8. Makihara, K.; Onoda, J.; Tsuchihashi, M. Investigation of Performance in Suppressing Various Vibrations with Energy-Recycling Semi-Active Method. *Acta Astronaut.* **2006**, *58*, 506–514. [CrossRef]
9. Zuo, L.; Zhang, P. Energy Harvesting, Ride Comfort, and Road Handling of Regenerative Vehicle Suspensions. *J. Vib. Acoust.* **2013**, *135*, 1060–1068. [CrossRef]
10. Zhang, Y.; Huang, K.; Yu, F.; Gu, Y.; Li, D. Experimental verification of energy-regenerative feasibility for an automotive electrical suspension system. In Proceedings of the 2007 IEEE International Conference on Vehicular Electronics and Safety, Beijing, China, 13–15 December 2007; pp. 13–15.
11. NHTS Administration. *Traffic Safety Facts*; National Highway Traffic Safety Administration: Washington, DC, USA, 2004. Available online: https://crashstats.nhtsa.dot.gov/Api/Public/ViewPublication/809919 (accessed on 10 May 2021).
12. NHTSA Administration. *Traffic Safety Facts 2014: A Compilation of Motor Vehicle Crash Data from the Fatality Analysis Reporting System and the General Estimates System*; National Highway Traffic Safety Administration: Washington, DC, USA, 2014.
13. Smith, W.A.; Zhang, N.; Hu, W. Hydraulically interconnected vehicle suspension: Handling performance. *Veh. Syst. Dyn.* **2011**, *49*, 87–106. [CrossRef]
14. Moulton, A.; Best, A. Hydragas®Suspension. In *1979 Automotive Engineering Congress and Exposition*; Paper Series 1979-01-790374; SAE: Detroit, MI, USA, 1979.

15. Wang, L.; Zhang, N.; Du, H. Experimental Investigation of a Hydraulically Interconnected Suspension in Vehicle Dynamics and Stability Control. *SAE Int. J. Passeng. Cars Mech. Syst.* **2012**, *5*, 759–768. [CrossRef]
16. Ding, F.; Han, X.; Luo, Z.; Zhang, N. Modelling and characteristic analysis of triaxle trucks with hydraulically interconnected suspensions. *Veh. Syst. Dyn.* **2012**, *50*, 1877–1904. [CrossRef]
17. Cao, D.; Rakheja, S.; Su, C. Property Analysis of an X-Coupled Suspension for Sport Utility Vehicles. *SAE Int. J. Passeng. Cars Mech. Syst.* **2008**, *1*, 853–862. [CrossRef]
18. Zhang, N.; Smith, W.A.; Jeyakumaran, J. Hydraulically interconnected vehicle suspension: Background and modelling. *Veh. Syst. Dyn.* **2010**, *48*, 17–40. [CrossRef]
19. Wang, M.; Zhang, B.; Chen, Y.; Zhang, N.; Zhang, J. Frequency-Based Modeling of a Vehicle Fitted with Roll-Plane Hydraulically Interconnected Suspension for Ride Comfort and Experimental Validation. *IEEE Access* **2020**, *8*, 1091–1104. [CrossRef]
20. Wang, L.; Jian, S.; Qi, H.; Zhang, N. Lateral stability study of a vehicle fitted with hydraulically interconnected suspension in slalom maneuver. In Proceedings of the 2017 Chinese Automation Congress (CAC), Jinan, China, 20–22 October 2017; pp. 1702–1707.
21. Zhang, D.; Li, H.; Dong, P.; Li, J.; Zhao, S. Study on a new damper for hydraulically interconnected suspension and its response characterization. In Proceedings of the CSAA/IET International Conference on Aircraft Utility Systems, Guiyang, China, 19–22 June 2018; pp. 394–398.
22. Wang, R.; Ye, Q.; Sun, Z.; Zhou, W.; Cao, Y.; Chen, L. A study of the hydraulically interconnected inerter-spring-damper suspension system. *Mech. Based Des. Struct. Mach.* **2017**, *45*, 415–429. [CrossRef]
23. Cao, D.; Subhash, R.; Su, Y. Roll- and pitch-plane-coupled hydro-pneumatic suspension. Part 2: Dynamic response analyses. *Veh. Syst. Dyn.* **2010**, *48*, 507–528. [CrossRef]
24. Wendel, G.R.; Stecklein, G.L. A regenerative active suspension system. In *International Congress & Exposition*; Paper Series 1991-01-910659; SAE: Detroit, MI, USA, 1991.
25. Liang, X.C.; Zhao, J.S. Research on Recycling Vibration Energy of Shock Absorber. *Int. J. Veh. Des.* **2015**, *68*, 201–220. [CrossRef]
26. Abdelkareem, M.; Xu, L.; Mohamed, K.A.A.; Elagouz, A.; Mi, J.; Guo, S.; Liu, Y.; Zuo, L. Vibration Energy Harvesting in Automotive Suspension System: A Detailed Review. *Appl. Energy* **2018**, *229*, 672–699. [CrossRef]
27. Zhao, Z.; Wang, T.; Zhang, B.; Shi, J. Energy Harvesting from Vehicle Suspension System by Piezoelectric Harvester. *Math. Probl. Eng.* **2019**, *2019*, 1086983. [CrossRef]
28. Wang, Z.; Zhang, T.; Zhang, Z.; Yuan, Y.; Liu, Y. A High-Efficiency Regenerative Shock Absorber Considering Twin Ball Screws Transmissions for Application in Range-Extended Electric Vehicles. *Energy Built Environ.* **2020**, *1*, 36–49. [CrossRef]
29. Li, Z.; Zuo, L.; Luhrs, G.; Lin, L.; Qin, Y. Electromagnetic Energy-Harvesting Shock Absorbers: Design, Modeling, and Road Tests. *IEEE Trans. Veh. Technol.* **2013**, *62*, 1065–1074. [CrossRef]
30. Lv, X.; Ji, Y.; Zhao, H.; Zhang, J.; Zhang, G.; Zhang, L. Research Review of a Vehicle Energy-Regenerative Suspension System. *Energies* **2020**, *13*, 441. [CrossRef]
31. Wu, Z.; Xu, G. Modeling and Analysis of a Hydraulic Energy Regeneration Shock Absorber. *Math. Probl. Eng.* **2020**, *2020*, 1580297.
32. Samn, A.; Abdelhaleem, A.; Kabeel, A.; Gad, E. Ride Comfort, Road Holding, and Energy Harvesting of a Hydraulic Regenerative Vehicle Suspension. *SAE Int. J. Passeng. Cars Mech. Syst.* **2020**, *13*, 159–171. [CrossRef]
33. Fang, Z.; Guo, X.; Xu, L. An optimal algorithm for energy recovery of hydraulic electromagnetic energy-regenerative shock absorber. *Appl. Math. Inf. Sci.* **2013**, *7*, 2207–2214. [CrossRef]
34. Guo, S.; Xu, L.; Liu, Y.; Guo, X.; Zuo, L. Modeling and Experiments of a Hydraulic Electromagnetic Energy Harvesting Shock Absorber. *IEEE Trans. Mechatron.* **2017**, *22*, 2684–2694. [CrossRef]
35. Zhou, Q.; Guo, S.; Xu, L.; Guo, X.; Yan, F.; Williams, H.; Xu, H. Global Optimization of the Hydraulic-electromagnetic Energy-harvesting Shock Absorber for Road Vehicles with Human-knowledge-integrated Particle Swarm Optimization Scheme. *IEEE Trans. Mechatron.* **2021**, 2, 1580297.
36. Peng, M.; Guo, X.; Zou, J.; Zhang, C. Simulation Study on Vehicle Road Performance with Hydraulic Electromagnetic Energy-Regenerative Shock Absorber. In *International Congress & Exposition*; Paper Series 2016-01-1550; SAE: Detroit, MI, USA, 2016.
37. Chen, Y.; Qin, B.; Guo, S.; Yu, L.; Zuo, L. Asymmetric Energy Harvesting and Hydraulically Interconnected Suspension: Modeling and Validations. In Proceedings of the ASME 2019 International Design Engineering Technical Conferences and Computers and Information in Engineering Conference, Anaheim, CA, USA, 18–21 August 2019; Volume 3, p. V003T01A009.
38. Guo, S.; Chen, Z.; Guo, X.; Zhou, Q. Vehicle Interconnected Suspension System based on Hydraulic Electromagnetic Energy Harvest: Design, Modeling and Simulation Tests. In *SAE 2014 Commercial Vehicle Engineering Congress*; Paper Series 2014-01-2299; SAE: Rosemont, IL, USA, 2014.
39. Zou, J.; Guo, X.; Abdelkareem, M.A.A.; Xu, L.; Zhang, J. Modelling and ride analysis of a hydraulic interconnected suspension based on the hydraulic energy regenerative shock absorbers. *Mech. Syst. Signal Process.* **2019**, *127*, 345–369. [CrossRef]
40. Watton, J.; Holford, K.M.; Surawattanawan, P. Electrohydraulic effects on the modelling of a vehicle active suspension. *Proc. Inst. Mech. Eng. Part D J. Automob. Eng.* **2001**, *215*, 1077–1092. [CrossRef]
41. Goodson, R.; Leonard, R. A survey of modelling techniques for fluid line transients. *ASME J. Basic Eng.* **1972**, *94*, 474–482. [CrossRef]
42. Stecki, J.; Davis, D. Fluid transmission lines—Distributed parameter models part 2: Comparison of models. *Proc. Inst. Mech. Eng. Part A Power Process Eng.* **1986**, *200*, 229–236. [CrossRef]

43. Stecki, J.; Davis, D. Fluid transmission lines—Distributed parameter models part 1: A review of the state of the art. *Proc. Inst. Mech. Eng. Part A Power Process Eng.* **1986**, *200*, 215–228. [CrossRef]
44. Yu, B.; Wang, Z.; Wang, G.; Zhao, J.; Zhou, L.; Zhao, J. Investigation of the suspension design and ride comfort of an electric mini off-road vehicle. *Adv. Mech. Eng.* **2019**, *11*, 1–10. [CrossRef]

micromachines

MDPI

Article

Stretchable Strain Sensor with Controllable Negative Resistance Sensitivity Coefficient Based on Patterned Carbon Nanotubes/Silicone Rubber Composites

Rong Dong [1] and Jianbing Xie [2,*]

[1] School of Mechatronic Engineering, Xi'an Technological University, Xi'an 710021, China; dongrong1981@163.com

[2] Key Laboratory of Micro/Nano Systems for Aerospace, Ministry of Education, Northwestern Polytechnical University, Xi'an 710072, China

* Correspondence: xiejb@nwpu.edu.cn; Tel.: +86-29-88460434

Abstract: In this paper, stretchable strain sensors with a controllable negative resistance sensitivity coefficient are firstly proposed. In order to realize the sensor with a negative resistance sensitivity coefficient, a stretchable stress sensor with sandwich structure is designed in this paper. Carbon nanotubes are added between two layers of silica gel. When the sensor is stretched, carbon nanotubes will be squeezed at the same time, so the sensor will show a resistance sensitivity coefficient that the resistance becomes smaller after stretching. First, nanomaterials are coated on soft elastomer, then a layer of silica gel is wrapped on the outside of the nanomaterials. In this way, similar to sandwich biscuits, a stretchable strain sensor with controllable negative resistance sensitivity coefficient has been obtained. Because the carbon nanotubes are wrapped between two layers of silica gel, when the silica gel is stretched, the carbon nanotubes will be squeezed longitudinally, which increases their density and resistance. Thus, a stretchable strain sensor with negative resistance sensitivity coefficient can be realized, and the resistivity can be controlled and adjusted from 12.7 Ω·m to 403.2 Ω·m. The sensor can be used for various tensile testing such as human motion monitoring, which can effectively expand the application range of conventional tensile strain sensor.

Keywords: strain sensor; negative resistance sensitivity coefficient; carbon nanotubes (CNTs)

Citation: Dong, R.; Xie, J. Stretchable Strain Sensor with Controllable Negative Resistance Sensitivity Coefficient Based on Patterned Carbon Nanotubes/Silicone Rubber Composites. *Micromachines* **2021**, *12*, 716. https://doi.org/10.3390/mi12060716

Academic Editor: Beatriz Jurado Sánchez

Received: 15 May 2021
Accepted: 17 June 2021
Published: 19 June 2021

1. Introduction

Wearable technology tends to be more sophisticated than hand-held technology because it can provide skin-mountable biofeedback and tracking of physiological function. In particular, stretchable and wearable strain sensors are needed for several potential applications including human motion and health detection, human-robot interaction, artificial skin, smart clothing, and so forth [1,2].

Stretchability and sensitivity are the key features of strain sensor which can be described by Strain (ε) and Gauge Factor (GF), respectively. In recent years, several types of stretchable strain sensors have been proposed by using nanomaterials coupled with flexible and stretchable elastomer. To improve stretchability, the conductive particles such as nanoparticles (NPs) [3–6], carbon nanotubes (CNTs) [7–10], silver nanowires (AgNWs) [11,12], and graphene [13,14] are typically coated on soft elastomer, such as PDMS [13], Ecoflex [12,15], silicone elastomer [16,17], rubber [18,19], dragon-skin elastomer [20] et al. In [15], super-stretchable, skin-mountable, and ultra-soft strain sensors are presented by using carbon nanotube percolation network-silicone rubber nanocomposite thin films, the stretchability can achieve 500%, but the GF is only 1–2.5. In [21], AuNWs–latex rubber nanocomposite are used to achieve a new type of sensor featured a GF of $\approx$ 9.9 and stretchability of >350%. In [22], a strain sensor with Ecoflex rubber elastic substrate and rGO/DI sensing liquids is designed to make the super-elasticity possible. The stretchability can achieve 400%, and the GF is 31.6.

It can be found from the above research status, the resistance sensitivity is all positive, so in order to extend the application range of the stretchable strain sensor, this paper presents a controllable negative resistance sensitivity coefficient stretchable strain sensor. This is the first time that a negative resistance sensitivity coefficient stretchable strain sensor is proposed.

In [23], a strain sensor based on the sandwich-like PDMS/CNTs/PDMS composite is proposed. The strain sensor not only presents a good optical transmittance, but also gains the ability to monitor both the subtle motions of facial expressions and the large motions of human joints. On this basis, we found that the sandwich-like structure can also obtain a negative resistance sensitivity coefficient, which can further expand the application range of the sensor and its specific application will be developed in the following research.

2. Principle and Design

The typical structural composition of stretchable strain sensors is shown in Figure 1, nanomaterials are coated on soft elastomer, the elastomer core is a rectangular structure, the length (L), width (W), and thickness (T) are shown in Figure 1a,b. A "dragon-skin" silica gel with maximum strain up to 900% is used for soft elastomer, and the nanomaterials are single-walled carbon nanotubes. At this point, the resistance of the sensor depends on resistivity of the powder material, particle size of powder, and especially compactness of the powder material. As a stretchable strain sensor, when the elastomer core is stretched, as shown in Figure 1c, part of the transverse connection in the powder material will be broken, and the compactness of the powder material is reduced, thereby causing the sensor resistance to increase. Therefore, the stretchable strain sensor prepared by this structure scheme only has a positive resistance sensitivity coefficient.

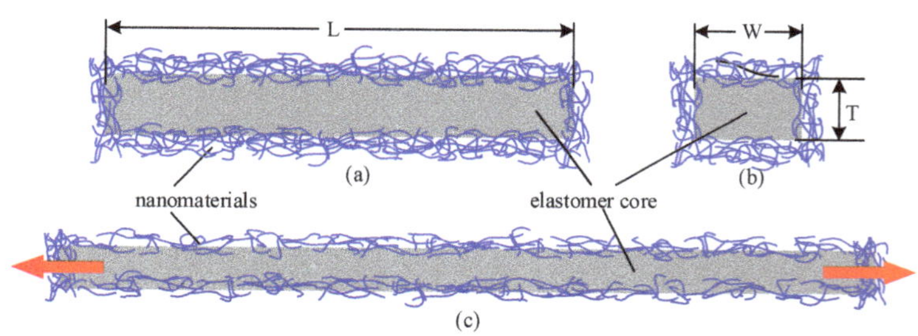

Figure 1. Typical structural composition of stretchable strain sensors, front view (**a**) and section view (**b**) before stretch and front view after stretch (**c**).

In order to extend the application range of the stretchable strain sensor, we present a controllable negative resistance sensitivity coefficient stretchable strain sensor, as shown in Figure 2. First, nanomaterials are coated on the soft elastomer. Then, a layer of silica gel is wrapped on the outside of the nanomaterials, as shown in Figure 2a. The thickness of the elastomer and nanomaterials is T_{ec} and T_{cnts}, respectively. When the elastomer is stretched, as shown in Figure 2b, although the lateral compactness of powder material is reduced, the longitudinal compactness increases.

Lateral compactness denoted the tightness of CNTs along the tensile direction. With the increase of tensile length, part of CNTs was disconnected, leading to the decrease of lateral compactness. Longitudinal compactness represents the compactness in the direction of the thickness of CNTs. With the increase of tensile length, CNTs have more contact under the longitudinal extrusion effect, which leads to the increase of longitudinal compactness.

Therefore, it is expected to obtain a stretchable strain sensor with a negative resistance sensitivity coefficient through reasonable pattern design.

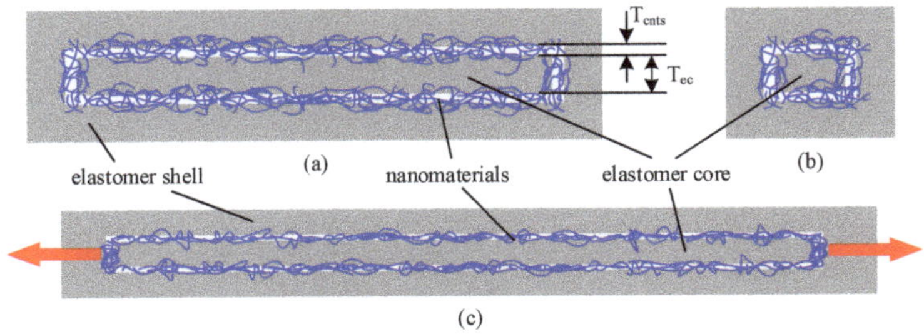

Figure 2. Controllable negative resistance sensitivity coefficient stretchable strain sensor, front view (**a**) and section view (**b**) before stretch and front view after stretch (**c**).

Normally, the powder resistivity measured by the pressurization method can be expressed as:

$$\rho_0 = \frac{V}{I}\frac{A}{h} \tag{1}$$

where ρ_0 is the initial resistivity of powder resistor, V is the applied voltage, I is the measured current, h is the powder thickness along the current direction, A is the cross-sectional area. It should be noted that powder resistivity varies a lot with powder porosity, humidity, or temperature.

Further, the resistance (R) of the stretchable strain sensor can be expressed as:

$$R = \rho\frac{L}{A} \tag{2}$$

where ρ is the resistivity of the sensor, L length of sensor. For the stretchable strain sensors designed in this paper, its resistivity, length, and cross-sectional area will change when stretched.

In order to verify the influence of different pattern structures on resistance, we designed several embedded resistance schemes as shown in Figure 3. Three stretchable strain sensors with different nanomaterial patterns are designed to verify the feasibility of the negative resistance sensitivity coefficient, which are defined as Type A, Type B, and Type C, respectively, as shown in Figure 3a–c, at the same time, typical stretchable strain sensors with the same patterns of Type A are designed as a comparison, which are defined as Type D, as shows in Figure 3d. The cross section of all resistors is rectangular. Since the unencapsulated structures corresponding to Type B and Type C will deform during stretch, it is impossible to compare the tensile effect. Therefore, this paper only carried out comparative experiments through Type A.

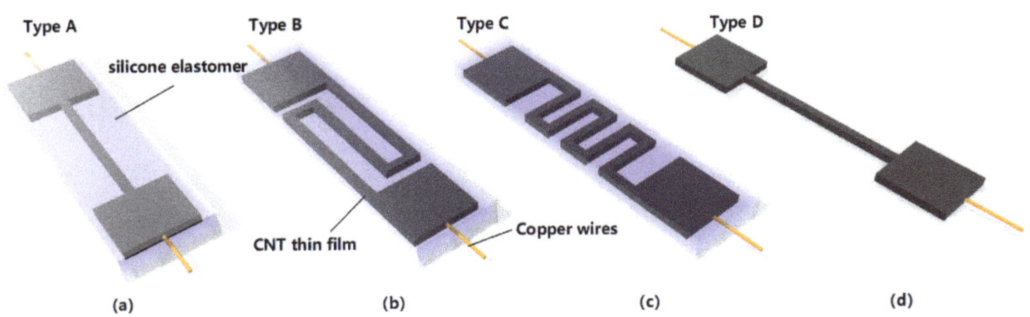

Figure 3. Designed controllable negative resistance sensitivity coefficient stretchable strain sensor (**a**–**c**) and typical structural of stretchable strain sensor (**d**).

3. Fabrication

The fabrication process of the designed negative resistance sensitivity coefficient stretchable strain sensor is shown in Figure 4. The elastomer core is made of a kind of "dragon-skin" silica gel which has a maximum strain up to 900%, so it can provide enough deformation for the designed stretchable strain sensor.

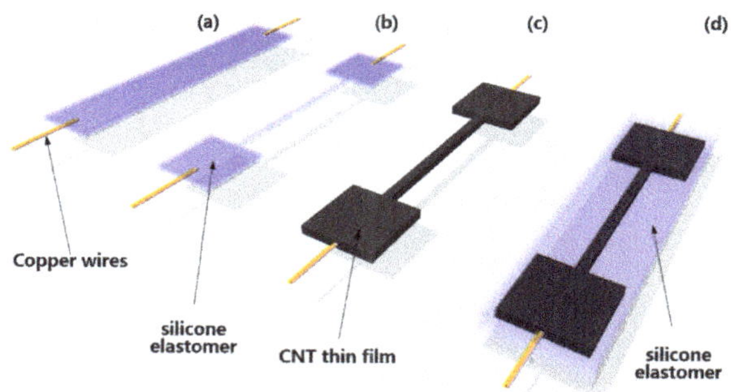

Figure 4. The fabrication process of the designed negative resistance sensitivity coefficient stretchable strain sensor. (**a**) Preparation of elastomer core and copper electrodes. (**b**) Silicone elastomer forming. (**c**) Preparation of CNTs film. (**d**) Silica gel secondary package.

First, insert two copper electrodes into the configured silica gel and then cure at room temperature (25 °C) for 1 h, as shown in Figure 4a. At this time, the surface of the silica gel has a good viscosity, but it has been substantially formed and can be cut into a desired shape, as shown in Figure 4b. The carbon nanotube powder is evenly dispersed on the surface of the silica gel elastomer core and compacted to obtain the stretchable strain sensor. At this time, after the further curing process, the typical structural of stretchable strain sensors can be obtained, as shown in Figure 4c.

At this stage of the process, we have the stretchable strain sensor of Type D, as shown in Figure 5a. This is a stretchable strain sensor with a positive resistance sensitivity.

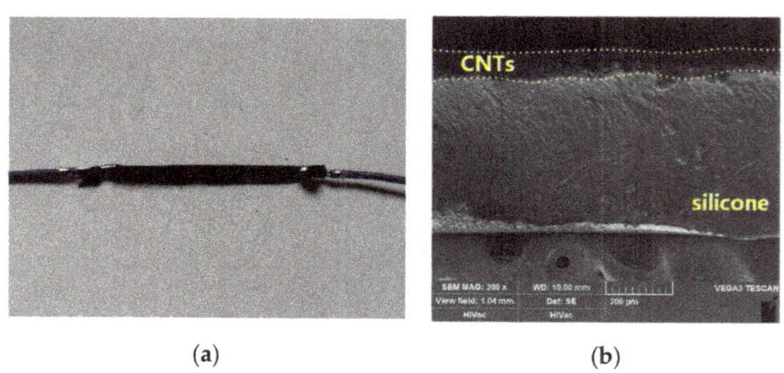

Figure 5. The fabricated stretchable strain sensor (**a**) and the profile SEM view (**b**).

However, the negative resistance sensitivity coefficient stretchable strain sensor proposed in this paper is re-encapsulated on this basis, as shown in Figure 4d. The overall structure shown in Figure 4c is encapsulated using the same "dragon-skin" silica gel, after that, the negative resistance sensitivity coefficient stretchable strain sensor can be obtained by curing at room temperature (25 °C) for 2 h.

The fabricated stretchable strain sensor (Type D) is shown in Figure 5. The thickness of the carbon nanotube layer is about 80 μm.

The fabricated stretchable strain sensors (Type A, Type B, and Type C) are shown in Figure 6. It can be seen that CNTs are well wrapped in silica gel and can be stretched together with silica gel.

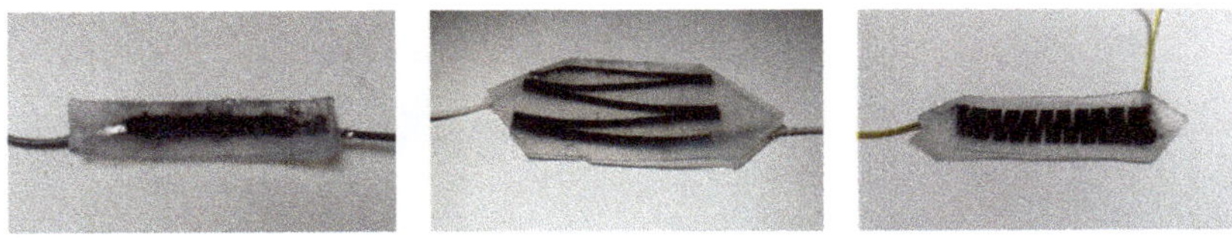

(a) (b) (c)

Figure 6. The fabricated stretchable strain sensor (**a**) Type A, (**b**) Type B, (**c**) Type C.

4. Characterization and Discussion

In order to verify the feasibility of the stretchable strain sensor with negative resistance sensitivity coefficient, a comparative test is carried out in this paper. The stretchable strain sensor is stretched to different lengths and its resistance is recorded. Figure 7 shows the stretch test of the fabricated stretchable strain sensor.

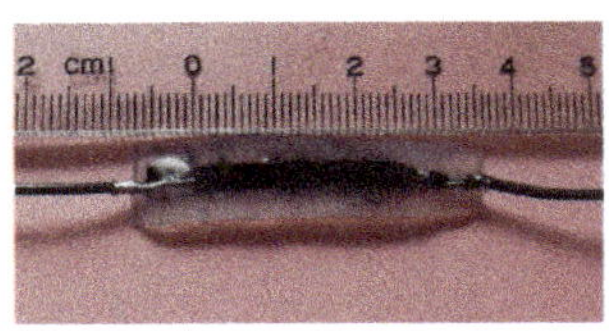

(a)

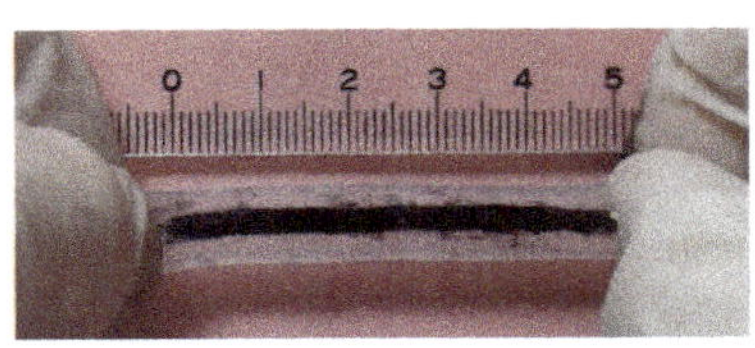

(b)

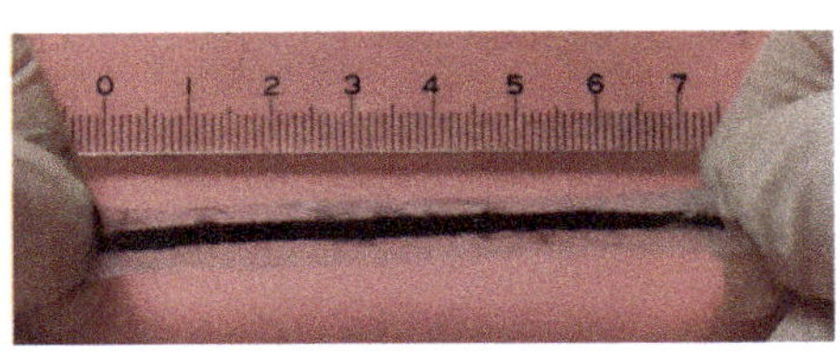

(c)

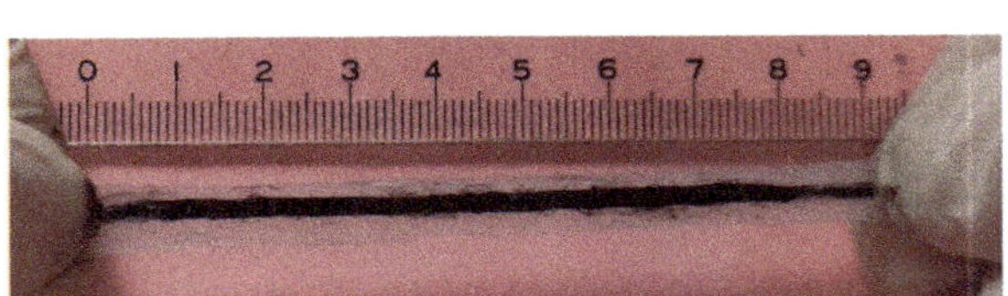

(d)

Figure 7. Stretch test of the fabricated stretchable strain sensor at $\varepsilon = 0\%$ (**a**), $\varepsilon = 100\%$ (**b**), $\varepsilon = 200\%$ (**c**), and $\varepsilon = 260\%$ (**d**).

The initial length of the sensor is 25 mm (Figure 7a), and it was stretched to 50 mm (Figure 7b), 75 mm (Figure 7c), and 90 mm (Figure 7d), respectively, to verify the resistance performance after stretching. It can also be seen from Figure 6 that the width of the resistor is basically unchanged, but the thickness gradually decreases with the stretching.

The resistance value and its change curve of the stretchable strain sensors after stretching are shown in Table 1. For the stretchable strain sensors with Type A, Type B, and Type C, as the stretched length of the sensor increases, the resistance increases continuously. However, for Type D, the resistivity increases rapidly with the increase of the stretch.

Table 1. Resistance test result of the stretchable strain sensors (The average of five measurements) @ 25 °C.

Sensor Type	Resistance @ ε = 0%(MΩ)	Resistance @ ε = 50%(MΩ)	Resistance @ ε = 100%(MΩ)	Resistance @ ε = 200%(MΩ)	Resistance @ ε = 260%(MΩ)
A	34.6 ± 12.4	27.4 ± 7.5	21.5 ± 6.9	14.2 ± 4.6	4.9 ± 1.8
B	194.5 ± 28.7	142.7 ± 18.7	105.6 ± 10.7	72.7 ± 8.7	36.8 ± 9.4
C	151.2 ± 20.4	120.4 ± 14.5	94.7 ± 10.2	52.4 ± 6.4	24.3 ± 7.5
D	1.35 ± 0.86	16.4 ± 4.7	42.2 ± 7.5	80.1 ± 10.1	126.2 ± 15.6

Figure 8 shows the SEM view of the fabricated stretchable strain sensors, before stretching, the thickness of the carbon nanotube layer is about 32 μm, as shown in Figure 8a, and after stretching the thickness decreased to 17.2 μm (ε = 100%) and 10.6 μm (ε = 200%), as shown in Figure 8b,c. This clearly shows that the thickness of the sensor is compressed as it is stretched, therefore, the change in resistance is no longer monotonically increasing.

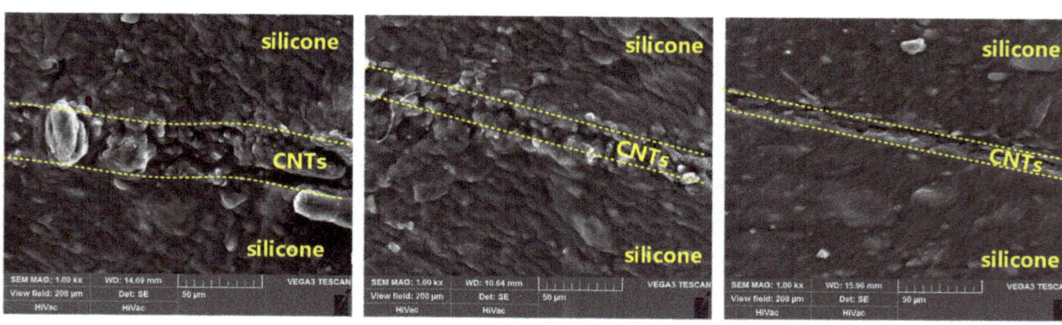

(a) (b) (c)

Figure 8. SEM view of the fabricated stretchable strain sensor at ε = 0% (**a**), ε = 100% (**b**) and ε = 200% (**c**).

According to the above experimental results, the resistivity of the sensor can be obtained, as shown in Table 2. It can be seen that for the stretchable strain sensors with Type A, Type B, and Type C, as the stretched length of the sensor increases, the resistivity increases continuously. This means that during the stretching process, the squeezing effect of the elastomer on the nanomaterial is greater than the stretching effect. When there is no external elastomer wrap (Type D), the resistivity increases rapidly with the increase of the stretch.

We can also see from Table 2 that the initial resistance between Type A and Type D is a lot different. This is because the liquid silica gel affected the contact of part of the carbon nanotubes when the Type D structure was repackaged into the Type A structure, leading to the decline of its initial resistance.

So, we can get the relationship curve between sensor length and resistivity, as shown in Figure 9. It can be seen that the stretchable strain sensors with Type A, Type B, and Type C have a negative resistance sensitivity coefficient.

As shown in Figure 9, the resistance sensitivity coefficient of Type D within ε = 100% is 12.23, by contrast, the resistance sensitivity coefficient of Type A within ε = 100% is negative 12.07 Ω·m/mm. At the same time, the resistance sensitivity coefficient can be controlled according to different lengths and patterns.

Table 2. Resistivity of the stretchable strain sensors @ 25 °C.

/	Type A			Type B			Type C			Type D		
Length (μm)	25	50	75	125	250	375	96	192	288	25	50	75
Resistance (MΩ)	34.6	21.5	14.2	194.5	105.6	72.7	151.2	94.7	52.4	1.35	42.2	80.1
Tec (mm)	1	0.55	0.3	8	7.1	6.6	8	7.1	6.6	8	7.1	6.6
Tcnts (μm)	32	17.2	10.6	32	17.2	10.6	32	17.2	10.6	55	55	55
Resistivity (Ω·m)	354.3	52.5	13.2	398.3	51.5	13.5	403.2	60.2	12.7	23.7	329.5	387.6

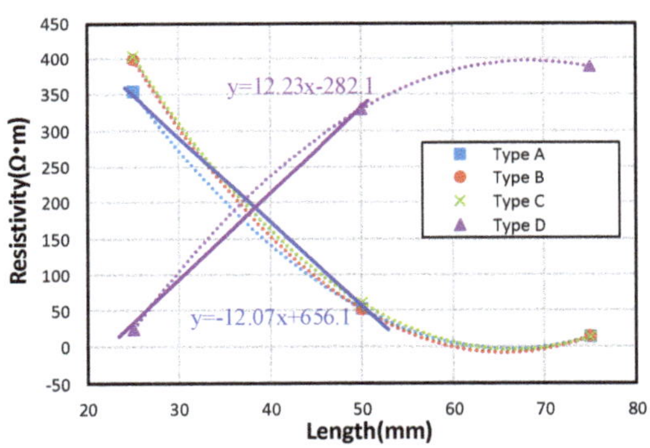

Figure 9. The relationship curve between sensor length and resistivity.

Therefore, the stretchable strain sensor with negative resistance sensitivity designed in this paper can expand the application of this device in many fields such as human motion monitoring.

5. Conclusions

This paper presents the design, fabrication of a novel stretchable strain sensors with a controllable negative resistance sensitivity coefficient. A stretchable stress sensor with silica gel/CNTs/silica gel sandwich structure is designed to realize the negative resistance sensitivity coefficient. The stretchability of the fabricated stretchable strain sensors can achieve up to 260%, resistance sensitivity coefficient is negative at 12.07 Ω·m/mm, and the adjustable control of resistivity from 354.3 Ω·m to 13.2 Ω·m, which can greatly expand the application field of the stretchable strain sensors such as human motion monitoring.

Author Contributions: Conceptualization, J.X.; methodology, R.D.; formal analysis, R.D.; data curation, R.D.; writing—original draft preparation, R.D.; writing—review and editing, J.X.; visualization, J.X.; supervision, J.X.; project administration, J.X.; funding acquisition, J.X. All authors have read and agreed to the published version of the manuscript.

Funding: This research was funded by National Natural Science Foundation of China Grant No., grant number 51775447.

Conflicts of Interest: The authors declare no conflict of interest.

References

1. Amjadi, M.; Kyung, K.-U.; Park, I.; Sitti, M. Stretchable, skin-mountable, and wearable strain sensors and their potential applications: A review. *Adv. Funct. Mater.* **2016**, *26*, 1678–1698. [CrossRef]
2. Tao, K.; Tang, L.; Wu, J.; Lye, S.W.; Chang, H.; Miao, J. Investigation of multimodal electret-based MEMS energy harvester with impact-induced nonlinearity. *J. Microelectromech. Syst.* **2018**, *27*, 276–288. [CrossRef]
3. Herrmann, J.; Muller, K.H.; Reda, T.; Baxter, G.; Raguse, B.; de Groot, G.; Chai, R.; Roberts, M.; Wieczorek, L. Na-noparticle films as Sensitive Strain Gauges. *Appl. Phys. Lett.* **2007**, *91*, 183105. [CrossRef]
4. Farcau, C.; Sangeetha, N.M.; Moreira, H.; Viallet, B.T.; Grisolia, J.; Ciuculescu-Pradines, D.; Ressier, L. High-sensitivity strain gauge based on a single wire of gold nanoparticles fabricated by stop-and-go con-vective self-assembly. *ACS Nano* **2011**, *5*, 7137–7143. [CrossRef] [PubMed]
5. Lee, J.; Kim, S.; Lee, J.; Yang, D.; Park, B.C.; Ryu, S.; Park, I. A stretchable strain sensor based on a metal nanoparti-cle thin film for human motion detection. *Nanoscale* **2014**, *6*, 11932. [CrossRef] [PubMed]
6. Kang, D.; Pikhitsa, P.V.; Choi, Y.W.; Lee, C.; Shin, S.S.; Piao, L.; Park, B.; Suh, K.-Y.; Kim, T.-I.; Choi, M. Ultrasensitive mechanical crack-based sensor inspired by the spider sensory system. *Nature* **2014**, *516*, 222–226. [CrossRef]
7. Lu, N.; Lu, C.; Yang, S.; Rogers, J. Highly sensitive skin-mountable strain gauges based entirely on elastomers. *Adv. Funct. Mater.* **2012**, *22*, 4044. [CrossRef]
8. Lipomi, D.J.; Vosgueritchian, M.; Tee, C.K.; Hellstrom, S.L.; Lee, J.A.; Fox, C.H.; Bao, Z. Skin-like pressure and strain sensors based on transparent elastic films of carbon nanotubes. *Nat. Nanotechnol.* **2011**, *6*, 788–792. [CrossRef]
9. Yamada, T.; Hayamizu, Y.; Yamamoto, Y.; Yomogida, Y.; Izadi-Najafabadi, A.; Futaba, D.N.; Hata, K. A stretchable carbon nanotube strain sensor for human-motion detection. *Nat. Nanotechnol.* **2011**, *6*, 296–301. [CrossRef]
10. Muth, J.T.; Vogt, D.M.; Truby, R.L.; Mengüç, Y.; Kolesky, D.B.; Wood, R.J.; Lewis, J.A. Embedded 3D printing of strain sensors within highly stretchable elastomers. *Adv. Mater.* **2014**, *26*, 6307–6312. [CrossRef]
11. Lee, S.; Shin, S.; Lee, S.; Seo, J.; Lee, J. Ag nanowire reinforced highly stretchable conductive fibers for wearable electronics. *Adv. Funct. Mater.* **2015**, *25*, 3114. [CrossRef]
12. Yao, S.; Zhu, Y. Wearable multifunctional sensors using printed stretchable conductors made of silver nan-owires. *Nanoscale* **2014**, *6*, 2345–2352. [CrossRef] [PubMed]
13. Lee, C.; Jug, L.; Meng, E. High strain biocompatible polydimethylsiloxane-based conductive graphene and multiwalled carbon nanotube nanocomposite strain sensors. *Appl. Phys. Lett.* **2013**, *102*, 183511. [CrossRef]
14. Hwang, S.-H.; Park, H.W.; Park, Y.-B. Piezoresistive behavior and multi-directional strain sensing ability of carbon nanotube–graphene nanoplatelet hybrid sheets. *Smart Mater. Struct.* **2013**, *22*, 015013. [CrossRef]
15. Amjadi, M.; Yoon, Y.J.; Park, I. Ultra-stretchable and skin-mountable strain sensors using carbon nano-tubes-Ecoflex nanocomposites. *Nanotechnology* **2015**, *26*, 375501. [CrossRef]
16. Cohen, D.J.; Mitra, D.; Peterson, K.; Maharbiz, M.M. A highly elastic, capacitive strain gauge based on perco-lating nanotube networks. *Nano Lett.* **2012**, *12*, 1821. [CrossRef]
17. Amjadi, M.; Pichitpajongkit, A.; Lee, S.; Ryu, S.; Park, I. Highly stretchable and sensitive strain sensor based on silver nanowire–elastomer nanocomposite. *ACS Nano* **2014**, *8*, 5154–5163. [CrossRef]
18. Boland, C.S.; Khan, U.; Backes, C.; O'Neill, A.; McCauley, J.; Duane, S.; Shanker, R.; Liu, Y.; Jurewicz, I.; Dalton, A.; et al. Sensitive, high-strain, high-rate bodily motion sensors based on graphene–rubber composites. *ACS Nano* **2014**, *8*, 8819–8830. [CrossRef]
19. Tang, Y.; Gong, S.; Chen, Y.; Yap, L.W.; Cheng, W. Manufacturable conducting rubber ambers and stretchable conductors from copper nanowire aerogel monoliths. *ACS Nano* **2014**, *8*, 5707–5714. [CrossRef]
20. Cai, L.; Song, L.; Luan, P.; Zhang, Q.; Zhang, N.; Gao, Q.; Zhao, D.; Zhang, X.; Tu, M.; Yang, F.; et al. Super-stretchable, transparent carbon nanotube-based capacitive strain sensors for human motion detection. *Sci. Rep.* **2013**, *3*, 3048. [CrossRef] [PubMed]
21. Gong, S.; Lai, D.T.H.; Su, B.; Si, K.J.; Ma, Z.; Yap, L.W.; Guo, P.; Cheng, W. Highly stretchy black gold e-skin nanopatches as highly sensitive wearable biomedical sensors. *Adv. Electron. Mater.* **2015**, *1*, 1400063. [CrossRef]
22. Xu, M.; Qi, J.; Li, F.; Zhang, Y. Highly stretchable strain sensors with reduced graphene oxide sensing liquids for wearable electronics. *Nanoscale* **2018**, *10*, 5264–5271. [CrossRef] [PubMed]
23. Chen, J.; Zhu, Y.; Jiang, W. A stretchable and transparent strain sensor based on sandwich-like PDMS/CNTs/PDMS composite containing an ultrathin conductive CNT layer. *Compos. Sci. Technol.* **2020**, *186*, 107938. [CrossRef]

micromachines

Article

A Microtester for Measuring the Reliability of Microdevices in Controlled Environmental Conditions

Yunjia Li [1,*], Weitao Dou [1], Chenyuan Zhou [1], Xinyi Wang [1], Aijun Yang [1], Yong Zhang [1] and Dayong Qiao [2,*]

1 School of Electrical Engineering, Xi'an Jiaotong University, Xi'an 710049, China; dou_wt@stu.xjtu.edu.cn (W.D.); zhouchenyuan@stu.xjtu.edu.cn (C.Z.); xywang@stu.xjtu.edu.cn (X.W.); yangaijun@xjtu.edu.cn (A.Y.); zhyong@mail.xjtu.edu.cn (Y.Z.)

2 Key Laboratory of Micro/Nano Systems for Aerospace, Ministry of Education, Northwestern Polytechnical University, Xi'an 710072, China

* Correspondence: liyunjia@xjtu.edu.cn (Y.L.); dyqiao@nwpu.edu.cn (D.Q.)

Abstract: A miniaturized reliability test system for microdevices with controlled environmental parameters is presented. The system is capable of measuring key electrical parameters of the microdevices while controlling the environmental conditions around the microdevices. The test system is compact and thus can be integrated with standard test equipment for microdevices. By using a feed-forward decoupling algorithm, the presented test system is capable of generating a temperature range of 0–120 °C and a humidity range of 20–90% RH (0–55 °C), within a small footprint and weight. The accuracy for temperature and humidity control is ±0.1 °C and ±1% RH (30 °C), respectively. The functionality of the proposed test system is verified by integrating it with a piezo shaker to test the environmental reliability of an electromagnetic vibration energy harvester. The proposed system can be used as a proof-of-technology platform for characterizing the performance of microdevices with controlled environmental parameters.

Keywords: temperature and humidity control; decoupling control; reliability test; microtester

Citation: Li, Y.; Dou, W.; Zhou, C.; Wang, X.; Yang, A.; Zhang, Y.; Qiao, D. A Microtester for Measuring the Reliability of Microdevices in Controlled Environmental Conditions. *Micromachines* **2021**, 12, 585. https://doi.org/10.3390/mi12050585

Academic Editor: Cheng Luo

Received: 25 April 2021
Accepted: 18 May 2021
Published: 20 May 2021

1. Introduction

Reliability of microelectromechanical system (MEMS) is one of the essential factors that determines the development time and time-to-market of the devices. Besides the design and the fabrication processes, environmental factors such as temperature, humidity and vibration are important aspects that influence the in-use reliability of MEMS devices [1]. In addition, environmental factors are known acceleration factors for accelerated reliability test of MEMS. For instance, temperature is often used to accelerate creep, electrical short/open-circuit, charging and corrosion failures, humidity is commonly used to accelerate cyclic fatigue, stiction and charging failures, and vibration is usually used to accelerate stiction and shock/vibration failures [1].

A great deal of work has been carried out to characterize MEMS reliability under different environmental conditions. For instance, Pustan et al. investigated the influence of temperature and humidity on the micromirror devices by using the combination of an Atomic Force Microscope and an environmental chamber [2]. It was discovered that the stiffness of the micromirror is reduced by 46% in the temperature range of 20 °C to 140 °C. Zhang et al. used the standalone environmental chamber to study the influence of temperature and humidity on the linearity and sensitivity of RF MEMS power sensors [3]. The test temperature and humidity ranges are 5 °C to 75 °C and 25% to 95% RH, respectively. Jan et al. proposed a test platform that combines a double-layer environmental chamber and a vibration shaker to study the effect of temperature and humidity on the resonant frequency of a CMOS-MEMS paddle resonator [4]. Resonant frequency drops of 6.9 Hz and 1.3 Hz were observed in the temperature range of 25–80 °C and humidity range of 32–90% RH, respectively. The vibration is asserted to the device through a hole on the bottom of the environmental chamber, which creates inhomogeneity of temperature distribution within the chamber. Sivakumar et al. characterized the accrual rate of stiction and mechanical fatigue in a MEMS micromirror device within a temperature and humidity range of 25 °C to 90 °C and 20% to 80% RH, respectively [5]. Lin et al. studied the RF MEMS switches performance and reliability under different temperatures [6]. Under temperature aging, the calculated Mean Time to Failure (MTTF) for each condition is 985 h, 822 h and 751 h for the temperatures of 50 °C, 85 °C and 105 °C, respectively.

Despite the great need of studying the MEMS reliability under different environmental conditions, one of the major challenges is to conduct those tests with different environmental factors using standard MEMS testing equipment, such as interferometer, laser-Doppler vibrometer, vibration table (piezo shaker), and different microscopes, etc. Some of the MEMS testing equipment is too bulky to be operated in the commercially available environmental chambers or ovens, while some other equipment is highly sensitive to environmental factors such as the piezo shakers. Taking piezo shaker as an example, the performances of piezoelectric materials are strongly influenced by temperature, especially around the Curie temperature of the material [7]. Therefore, it is difficult to use the piezo shaker within an oven. Meanwhile, most of the piezo shakers can generate very limited force, often in the range of tens of Newton, thus it is equivalently difficult to mount an oven, even the lightest one, on the piezo shaker. Such dilemma might be the detrimental factor for the lack of reliability of vibration energy harvesters (VEHs), in spite of a great deal of effort in developing various novel VEH devices [8–11].

Therefore, there is a strong need and market pull for standard generators of different environmental condition (environmental chambers), which can be integrated on different standard MEMS test equipment. For this purpose, the environmental chamber must be compact, lightweight and capable of generating a wide range of environmental conditions. Towards these goals, the major technical challenge is to integrate all functional elements within a rather small volume and not to exceed the mass and volume limitation. In addition, with miniaturized heating and humidifying elements, it is not easy to realize heating and humidifying functions with reasonable range and speed, compared to their macroscopic counterparts. Although control algorithms can be utilized to optimize the heating and humidifying process, most of the reported algorithms are developed for standard "big" environmental chambers [12–14]. Their applicability on miniaturized systems still need to be investigated.

In this paper, we proposed a miniaturized reliability test system for microdevices with controlled environmental parameters, which has the advantages of lightweight, small volume, wide and stable range of controllable environmental parameters. The system can be integrated with the piezo shaker, white light interferometer and other equipment to complete the performance characterization of microdevices under different environmental conditions. The technical merits of the proposed system are to integrate all functional components within a very limited space, and fulfill all heating, humidifying and measurement functionalities comparable to bulky environmental systems. Although the test system in this paper is a specific system, it provides a technical platform for the reliability test of microdevices and can be widely used.

2. System Design

2.1. Overal System Design

The concept of the proposed test system is to control the environmental parameters around the Device Under Test (DUT) while other performance of the DUT is tested. This requires the test chamber to be highly miniaturized and lightweight, in order to be integrated with other characterization tools for microdevices. It is especially crucial for the reliability characterization of Vibration Energy Harvesters (VEHs), because the test chamber needs to be mounted on the piezo shaker, whose loading capability is strongly limited in terms of pushing force. Objects with large mass will deteriorate the performance of the shaker significantly.

The proposed reliability test system is a miniaturized and lightweight equipment, based on a structure illustrated schematically in Figure 1. The system is composed of two main components: a test chamber and a control unit, which are connected by electrical cables and tubes. The test chamber is a miniaturized metallic chamber capable of changing its internal environmental parameters, mainly temperature and humidity, accurately and rapidly. The test chamber is highly compact and lightweight, equipped with electrical feedthrough and optical observation window, enabling its integration to standard characterization equipment such as white light interferometer, laser Doppler vibrometer and piezo shaker. The control unit includes a humidity regulator, temperature and humidity control circuits, electrical measurement circuits and a human–computer interaction module. The electrical measurement circuits include mainly components for capacitive and resistive measurements, and will not be detailed in this paper.

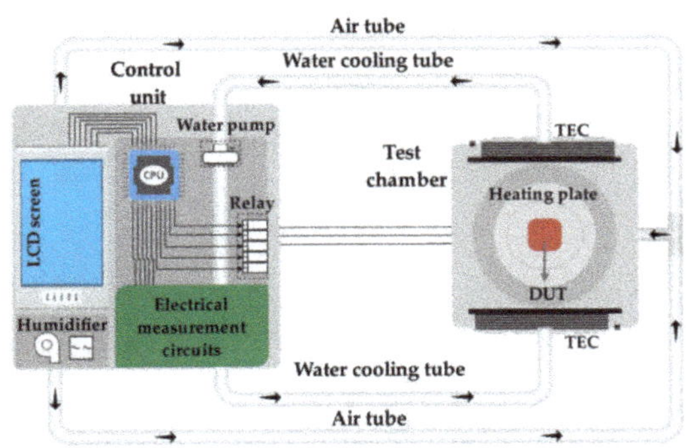

Figure 1. Schematic illustration of the miniaturized reliability test system with controlled environment parameters.

More specifically, the temperature regulator consists of a ceramic heating plate with a power of 96 W and two heat-resistant thermoelectric coolers (TECs) based on Peltier effect with power of 36 W. To achieve a better cooling while maintaining a designated volume of the test chamber, water cooling is used to dissipate heat from the hot end of the TEC. The humidity in the test chamber is controlled by using an ultrasonic transducer. The fan transmits the water vapor with a certain controlled humidity to the test chamber through the plastic tube. The interaction module consists of an LCD display screen and a STM32 microprocessor. The LCD screen displays the temperature and humidity data in the test chamber as well as the capacitance and resistance measured by the measurement module. By touching the LCD screen, control commands are sent to the temperature and humidity control module, which controls the ceramic heating plate, TECs and humidifier to change the temperature and humidity in the test chamber.

2.2. Test Chamber Design

In order to achieve maximum heating and humidifying capacity with minimum size and weight, the test chamber must be carefully designed, and elements and their layout within the chamber must be optimized. A schematic illustration of the designed test chamber is shown in Figure 2a, and an exploded view of the test chamber is detailed in Figure 2b. In order to maximize the efficiency of the system, the temperature regulator is directly installed within the test chamber. The ceramic heating plate is sandwiched between two sample stages, and the TECs are installed on both sides of the test chamber. A temperature and humidity sensor (Sensirion SHT31) is installed on the bottom side of the sample stage to monitor the real-time temperature and humidity in the closed vicinity of DUT. During measurements, the DUT is mounted on the top surface of the sample stage, with several mounting and pin-configuration possibilities enabled by pre-defined mounting plugs. For the heating processes, the sample stage will quickly transfer heat between the ceramic heating plate and the DUT. For the cooling processes, the hot end and the cold end of TEC will release heat and absorb heat respectively, due to the Peltier effect. Heat sink 2 mounted on the cold end of the TECs is used to quickly transfer heat between the cold end of the TEC and the chamber, while heat sink 1 mounted on the hot side of the TEC removes heat from the hot end of the TEC through water cooling. The combination of the passive and active water-cooled heatsinks enabled maximum efficiency and stability for the cooling processes. The mounting sites of the TECs and heatsinks are chosen under the criteria of shortest distance to the DUT. For the humidity control processes, water vapor with a controlled humidity enters the test chamber through the tube connector at the back of the chamber. The humidity in the water vapor is adjusted by using an ultrasonic transducer installed within a water tank inside the control unit. The entire model design is based on a symmetrical rule to ensure the homogeneity of temperature and humidity within the test chamber.

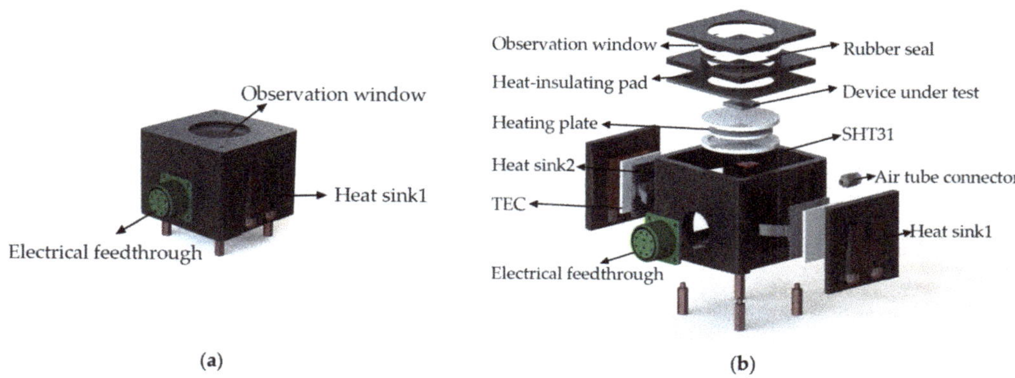

Figure 2. (**a**) Schematic illustration of the assembled test chamber, (**b**) schematic illustration of the exploded structure of the test chamber.

2.3. Control Unit Design

The control unit of the test system includes a power management circuit, an MCU circuit, a relay circuit, a sensing circuit and the electrical measurement circuits. Due to the complexity of the entire circuit system, a hierarchical design approach is adopted to construct the system from modular components with logic orders. A schematic illustration of the overall circuit is shown in Figure 3. The power management circuit provides power with different voltage levels to the various components in the system with different power requirements. The MCU circuit is used to realize all the designated functions of the STM32 microprocessor, including executing, uploading and downloading programs, implementing all the calculation and control algorithms, as well as communicating with other components. The relay circuit isolates the low-voltage side from the high-voltage side of the circuits by using an optocoupler, and controls the operation of the temperature and humidity regulator by using a high-power MOSFET as a switch. The sensor circuit drives the SHT31 temperature and humidity sensor to measure the real-time temperature and humidity data within the test chamber. The electrical measurement circuits measure the capacitance, resistance and other physical parameters of the DUT. In the circuit layout, the circuits are partitioned according to different voltage levels and functions. At the same time, single-point grounding of both the analog ground and digital ground is implemented to isolate the analog circuit from the digital circuit, avoiding the crosstalk between them. These design and structural concepts guaranteed the accuracy and reliability of the proposed test system.

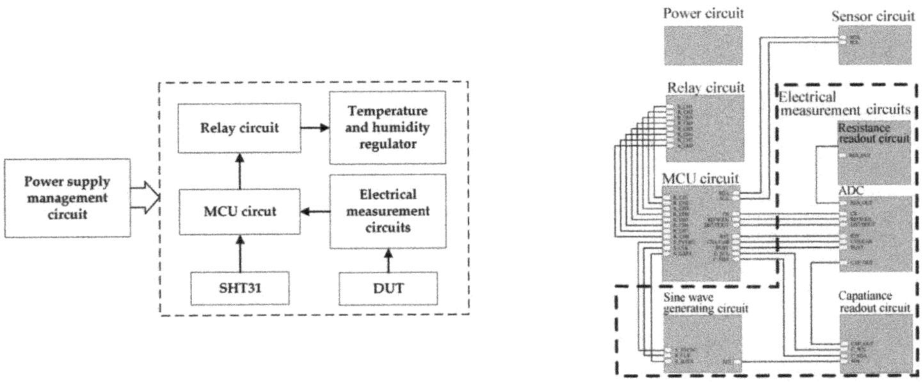

(a) (b)

Figure 3. (**a**) Block diagram of the control circuit, (**b**) schematic illustration of the control circuit.

2.4. Control Algorithms

Rapid control of temperature and humidity simultaneously under limitations of mass and volume is challenging, due to limited choice of components, strong temperature/humidity coupling, performance hysteresis and nonlinear dynamic characteristics. Consequently, a simple PID controller will not be sufficient to control the system [14,15]. Therefore, the proposed test system utilizes a feed-forward decoupling control algorithm to decouple and control the temperature and humidity, improving the accuracy of the system [16]. The feed-forward decoupling control algorithm is used to predict the effect of the coupling between the temperature and humidity on the control process, and subsequently eliminate the cross-coupling effect. The block diagram of the heating-humidifying decoupling controller is shown in Figure 4a. Similarly, the cooling-humidifying decoupling control block diagram is obtained, as shown in Figure 4b.

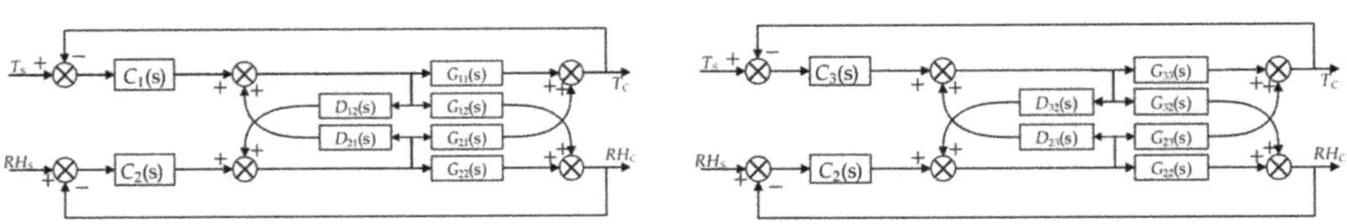

(a) (b)

Figure 4. (**a**) The heating-humidifying decoupling control block diagram, (**b**) the cooling-humidifying decoupling control block diagram.

It is crucial for the feed-forward decoupling control system to solve the feed-forward decoupling transfer function. The output of the heating-humidifying decoupling system can be described as:

$$[T_C \; RH_C] = \begin{bmatrix} C_1(s) & C_2(s) \end{bmatrix} \times \begin{bmatrix} G_{11}(s) & G_{12}(s) + G_{22}(s) \times D_{12}(s) \\ G_{21}(s) + G_{11}(s) \times D_{21}(s) & G_{22}(s) \end{bmatrix} \quad (1)$$

where $G_{11}(s)$ is the system heating model and $G_{12}(s)$ is the system heating-humidifying coupling model. $G_{21}(s)$ is the system humidifying-heating coupling model, $G_{22}(s)$ is the system humidifying model, $D_{21}(s)$ is the humidifying-heating feed-forward decoupling factor and $D_{12}(s)$ is the heating-humidifying feed-forward decoupling factor.

After the decoupling the heating and humidifying process, the ideal system output should be described as:

$$[T_C\ RH_C] = \begin{bmatrix} C_1(s) & C_2(s) \end{bmatrix} \times \begin{bmatrix} G_{11}(s) & 0 \\ 0 & G_{22}(s) \end{bmatrix} \tag{2}$$

From Equations (1) and (2), the relationship between the heating, humidifying feed-forward decoupling factor and the inherent model of the system can be obtained:

$$\begin{bmatrix} D_{21}(s) & 1 \\ 1 & D_{12}(s) \end{bmatrix} \times \begin{bmatrix} G_{11}(s) & G_{12}(s) \\ G_{21}(s) & G_{22}(s) \end{bmatrix} = \begin{bmatrix} 0 & 0 \\ 0 & 0 \end{bmatrix} \tag{3}$$

Similarly, the expression between the feed-forward decoupling factor of cooling, humidifying and the inherent model of the system can be obtained:

$$\begin{bmatrix} D_{32}(s) & 1 \\ 1 & D_{23}(s) \end{bmatrix} \times \begin{bmatrix} G_{22}(s) & G_{23}(s) \\ G_{32}(s) & G_{33}(s) \end{bmatrix} = \begin{bmatrix} 0 & 0 \\ 0 & 0 \end{bmatrix} \tag{4}$$

where, $G_{23}(s)$ is humidifying-cooling coupling model, $G_{32}(s)$ is cooling-humidifying coupling model, $G_{33}(s)$ is the system heating model, $D_{23}(s)$ is humidifying-cooling feed-forward decoupling factor and $D_{32}(s)$ is cooling-humidifying feed-forward decoupling factor.

2.5. Simulink Simulation

To verify the effectiveness of the control methodology, the control system is simulated by using Simulink according to the proposed control block diagram in Figure 5.

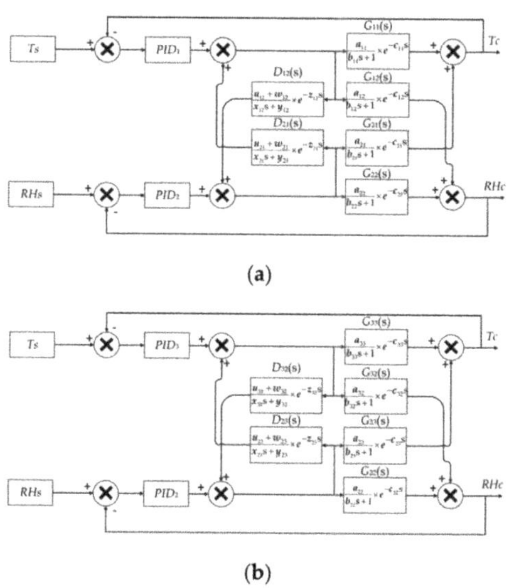

Figure 5. (**a**) The heating-humidifying system model, (**b**) the cooling-humidifying system model [16].

The model of the control system is effective in tuning the parameters of the controller, including both the feedback and feed-forward parameters, which are essential factors for a rapid and accurate control of temperature and humidity. In the actual tuning process, the parameters *a*, *b*, *c* of $G_{11}(s)$, $G_{12}(s)$, $G_{21}(s)$, $G_{23}(s)$ and $G_{33}(s)$ can be obtained experimentally [17], and the parameters *u*, *w*, *x*, *y*, *z* of feed-forward decoupling factors $D_{21}(s)$, $D_{12}(s)$, $D_{23}(s)$ and $D_{32}(s)$ can be solved by Equations (3) and (4). The solving process will be used for the tuning process detailed in Section 3.2. By implementing the discretized feed-forward decoupling factor differential equation into the control program, the effects of the system caused by the temperature and humidity coupling can be predicted and eliminated [18].

3. Results and Discussion

3.1. Manufacturing Results

The photos of the manufactured circuit board, control unit and test chamber are shown in Figure 6. The temperature and humidity controllers are connected to the KF-2.54 terminal in the relay circuit. To drive multiple high-power devices at the same time, the PCBs in the system involve design with windowed topology and heat dissipation holes, in order to increase the heat dissipation capacity of the circuit board. The control unit is fabricated by 3D printing technology, and the upper cover is equipped with switch buttons, control buttons, an interaction interface, and a water inlet for the humidity regulator. An optical observation window is installed on the top surface of the test chamber to enable the optical measurements on the DUT. The housing of the test chamber is made of aluminum alloy with surface treatment of thick anodization, in order to enhance corrosion resistance. The test chamber has external dimensions of 9.2 cm × 9.2 cm × 7.2 cm and a weight of 1.07 kg. The control unit and the test chamber are connected through high-temperature aviation plugs for reliable electrical connections.

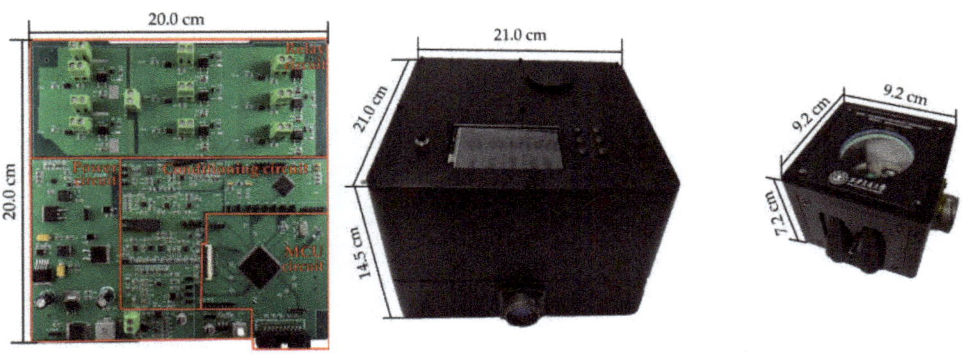

(a) (b) (c)

Figure 6. Photos of (**a**) the circuit board, (**b**) the control chamber, (**c**) the test chamber.

3.2. System Tuning

In order to solve the temperature and humidity model of the test chamber, the temperature and humidity open-loop response and the temperature and humidity open-loop coupling response of the test chamber are studied in time domain [19]. Figure 7a, b show the temperature open-loop response of the test system to a 50% PWM duty-cycle step cooling signal and humidity open-loop coupling response of the test system to a 30% PWM duty-cycle step cooling signal, respectively (PWM cycle is 1 s). The 50% PWM duty-cycle step cooling signal is given at 0 s with a rise time of 20 ms. The temperature change of the test chamber delay is 44 s (warming up of the cooler), then rapidly from 44 s to 250 s (cooling process), and finally gradually between 250 s and 450 s. At the initial temperature of 20 °C, the maximum cooling rate is 0.075 °C per second at 80 s, and the temperature of the test chamber is finally stabilized at 9.3 °C under a 50% PWM duty-cycle step cooling signal. The open-loop coupling response of the humidity within the chamber follows a similar pattern. The 30% PWM duty-cycle step cooling signal is given at 0 s with a rise time of 20 ms. The humidity change of the test chamber delay is 28 s, then rapidly from 28 s to 200 s, and finally gradually between 200 s and 350 s. At the initial humidity of 76.8% RH, the maximum dehumidifying rate is 0.095% RH per second at 50 s, and the humidity of the test chamber is finally stabilized at 63.3% RH under a 30% PWM duty-cycle step cooling signal. The coupling of cooling and humidity is mainly caused by the sudden drop of temperature at the TEC cold end, which results in condensation of water vapor in the air. Similarly, we can get the open-loop response curve and open-loop coupling response curve of heating and humidifying, which will not be repeated here.

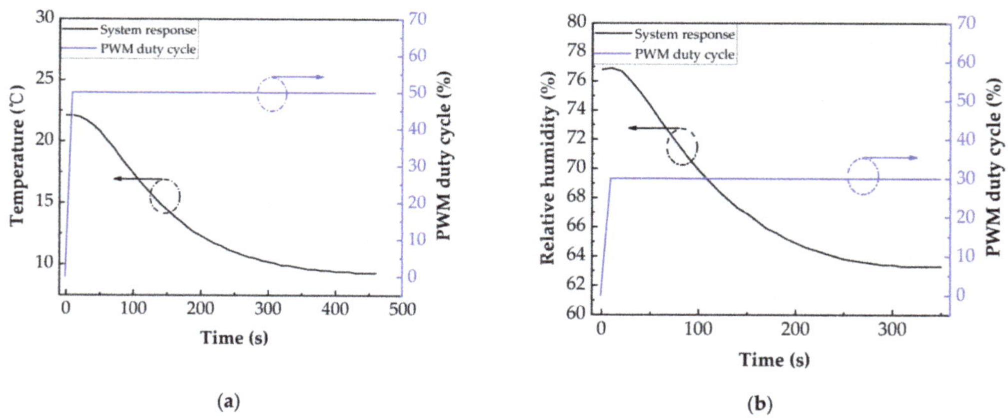

Figure 7. (**a**) Temperature response of the test system to a 50% PWM duty-cycle step cooling signal. (**b**) Relative humidity response of the test system to a 30% PWM duty-cycle step cooling signal.

The two-point method was used to obtain the parameters of first-order delay models of the system, as given in Table 1.

Table 1. Parameters of first-order delay models of the system.

Parameters	Values	Parameters	Values	Parameters	Values
a_{11}	0.1559	b_{11}	751.27	c_{11}	5
a_{12}	0.3357	b_{12}	1618.00	c_{12}	136
a_{21}	0.0004	b_{21}	428.45	c_{21}	100
a_{22}	0.1949	b_{22}	183.65	c_{22}	13
a_{23}	0.0050	b_{23}	976.46	c_{23}	105
a_{32}	0.0488	b_{32}	107.60	c_{32}	28
a_{33}	0.0270	b_{33}	124.80	c_{33}	44

Through Equations (3) and (4), the parameters of the transfer functions of feed-forward decoupling factor are derived, as shown in Table 2.

Table 2. Parameters of the transfer functions of feed-forward decoupling factor.

Parameters	Values	Parameters	Values	Parameters	Values	Parameters	Values	Parameters	Values
u_{12}	−61.6500	w_{12}	−0.3357	x_{12}	315.30	y_{12}	0.1949	z_{12}	135
u_{21}	−0.7036	w_{21}	−0.0009	x_{21}	1599.00	y_{21}	0.3639	z_{21}	95
u_{23}	0.6205	w_{23}	−0.0050	x_{23}	26.36	y_{23}	0.0270	z_{23}	65
u_{32}	−8.9690	w_{32}	−0.0488	x_{32}	20.97	y_{32}	0.1949	z_{32}	135

These parameters of transfer functions of the proposed system can be used to model the response of the control system. The analytical form of the model enabled the modeling process to be simple and accurate, where key design parameters can be tuned readily and conveniently.

The PID Tuner is utilized to solve the PID parameters in the system. The comparison between the system responses before and after the introduction of the feed-forward decoupling factor is shown in Figure 8. The target temperature of 70 °C is realized within the time of 400 s in Figure 8a and the target temperature of 5 °C is realized within the time of 200 s in Figure 8b, both showing no overshoot or oscillation behavior. The temperature response curves before and after decoupling in Figure 8a, b almost coincide, which means the feed-forward decoupling factor has almost no effect on the temperature regulation of the system. However, the feed-forward decoupling factor can obviously improve the humidity control of the system. In the heating-humidifying system of Figure 8a, the target humidity of 90% RH is realized within the time of 100 s, showing no overshoot or oscillation behavior. However, when the system temperature reaches 45.6 °C at 146 s, the humidity of the system before decoupling decreases at the rate of 0.004% RH per second until 420 s, and then increases to 88.7% RH at 1000 s. Compared with before decoupling, the humidity of the system after decoupling is always stable at a target humidity of 90% RH. In the cooling-humidifying system in Figure 8b, by introduction of the feed-forward decoupling factor, the regulating time decreases from 323 s to 108 s. To sum up, the simulation results show that the feed-forward decoupling factor has almost no effect on the temperature regulation of the system, but can obviously improve the humidity regulation.

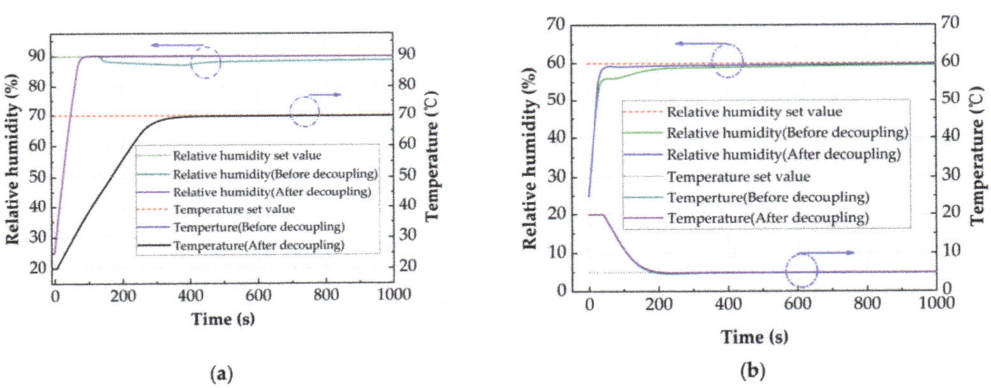

Figure 8. (**a**) Response curve of heating-humidifying system before and after decoupling. (**b**) Response curve of cooling-humidifying system before and after decoupling.

3.3. System Characterization

From the simulation in the previous section, it can be seen that humidification has little effect on refrigeration and heating, so only feed-forward decoupling factors $D_{12}(s)$ and $D_{32}(s)$ are introduced in the system. In order to introduce the feed-forward decoupling factor into the control system, we need to carry out Z transformation on the feed-forward decoupling factors, and then they are transformed into difference equations, finally.

By introducing the difference equations into the control program, the feed-forward decoupling factor of the actual control system is added. The PID parameters obtained from the simulation are brought into the actual system, and then fine-tuning is carried out to obtain the temperature control effect of the system shown in Figure 9a and the temperature and humidity decoupling control effect shown in Figure 9b.

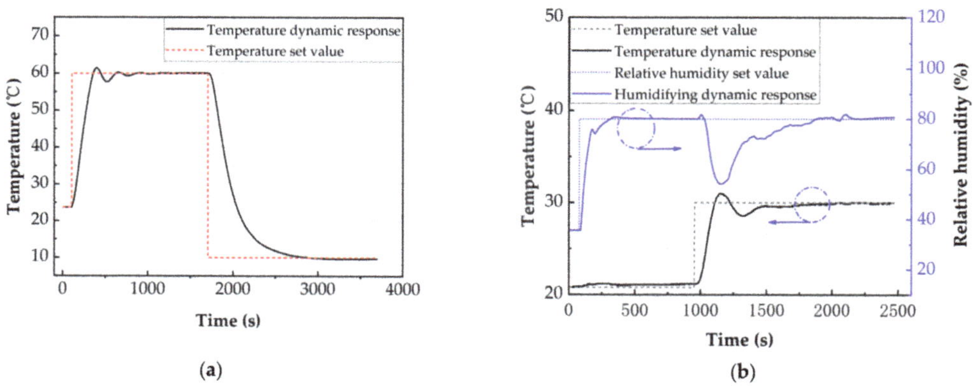

Figure 9. (**a**) System temperature dynamic response. (**b**) System temperature-humidity coupling dynamic response.

In order to characterize the heating dynamic adjustment performance of the system, a step temperature signal that has a target temperature of 60 °C is given at 100 s, as shown in Figure 9a. The target temperature of 60 °C is realized within the time of 2023 s and the maximum overshoot is 2.5%. The system temperature differs from the target temperature by 0.1 °C within 700 s, and gradually stabilizes at 60 °C. The characterization of the cooling dynamic adjustment performance of the system within the chamber follows a similar pattern. A step temperature signal that has a target temperature of 10 °C is given at 1706 s, as shown in Figure 9a. The temperature of the test chamber decreases rapidly from 1706 s to 2300 s (cooling process), and finally gradually between 2300 s and 2700 s (precise temperature control to avoid overshoot). The target temperature of 10 °C is realized within the time of 1080 s, showing no overshoot or oscillation behavior. The system temperature differs from the target temperature by 0.1 °C within 1124 s. The maximum temperature change rate is 0.2 °C/s for the heating process and −0.175 °C/s for the cooling process. Compared to the simulation, the actual temperature dynamic response of the system is slower because of the simplified first-order heating and cooling model, but can meet the requirements of most test devices.

The anti-disturbance test of the system is shown in Figure 9b, and a step humidity signal that has a target humidity of 80% RH is given at 176 s. The target humidity of 80% RH is realized within the time of 234 s, and the maximum overshoot is 1%. The system humidity differs from the target humidity by 0.2% RH within 474 s. During the humidification process, the system temperature first gradually rises in 40–308 s, and stabilizes at 21.2 °C in 380–410 s, and then gradually returns to the initial value. The reason for this phenomenon is the temperature of the atomizing gas will be slightly higher than the temperature in the test chamber. A step temperature signal that has a target temperature of 30 °C is given at 1054 s. Due to the feedforward decoupling factor, the system humidity will increase in advance before being affected by temperature. Therefore, the system humidity will be higher than the set value of 80% RH in 1074–1126 s, and then as the temperature continues to rise to 31 °C, the system humidity gradually decreases to 54.3% RH. As the system temperature gradually stabilizes, the system humidity is also slowly increasing, and finally stabilizes at 80% ± 1% RH at 2182 s. Compared to the simulation, during the system temperature increase, the internal temperature of the test chamber does not reach complete consistency, which causes the system humidity to fail to maintain the set value.

In order to study the static tuning range of the system, the humidity is set to 100% RH and the temperature is increased from 0 °C to 120 °C, with a step of 10 °C. When the system is in steady state, the maximum humidity of the system is recorded. As a next step, the humidity is set to 0% RH and the temperature is decreased from 120 °C to 0 °C, with a step of 10 °C. When the system is in steady state, the minimum humidity of the system is recorded. The measured static tuning range of the system is shown in Figure 10. In the temperature range of 0–40 °C, the maximum humidity that can be reached in the test chamber is 97% RH. The maximum humidity tuning range of 4.8–97% RH in the chamber is realized at a temperature of 40 °C. At elevated temperature, the maximum humidity within the chamber decreases. The measured maximum humidity is 70.3% RH at 60 °C, 40% RH at 90 °C and 9.3% RH at 120 °C.

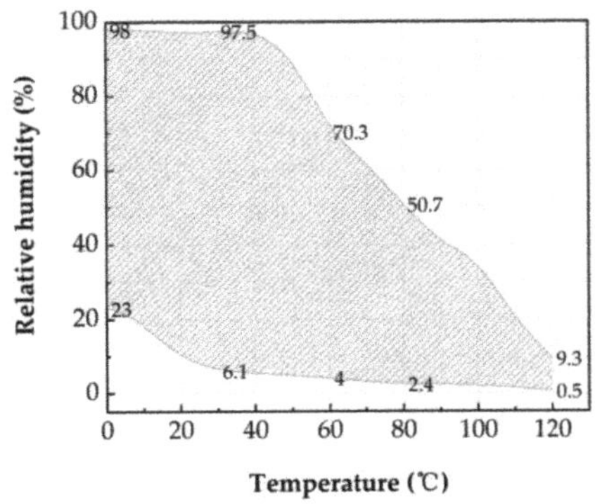

Figure 10. Relative humidity tuning characteristics in the test chamber as a function of temperature.

The maximum humidity in the chamber decreases rapidly when the temperature increases, which can be attributed to the following reasons: As the temperature increases, the saturated vapor is capable of holding more liquid content. The maximum level of liquid content in the vapor generated by the humidifier (held at room temperature) cannot exceed the level of liquid content in the saturated vapor. The minimum humidity of the system is 0.5% RH at 120 °C and 23% RH at 0 °C, which because of the moisture content in saturated air will decrease as the temperature decreases. Considering the insufficiencies above, there is still space for improvements in the humidification system. For example, different humidification devices can be used at different temperature ranges, i.e., ultrasonic atomization device can be used to humidify the chamber at a low-temperature range, and thermal vaporization humidifier at a high-temperature range.

3.4. System Functionality Verification

Figure 11a shows a photo of the miniaturized reliability test system integrated with a piezo shaker. Such experimental configuration enables the reliability test of micro-energy harvesters, which have not been reported due to the incompatibility between the piezo shaker and environmental chambers (e.g., a convection oven). The shaker used in this study is a model JZK-20 piezo shaker from Sinocera Corporation, which is capable of generating a maximum excitation force of 200 N. The DUT is mounted on a customized sample stage with electrical connections to the feedthrough on the sidewall of the chamber, as shown in Figure 11b. The sample stage can be readily exchanged to adapt to different types of device connections. A second temperature sensor is mounted above the sample stage next to the DUT in order to monitor the actual temperature around the DUT. The DUT is a customized electromagnetic vibration energy harvester fabricated by the authors for the verification of the proposed test system. It is based on a flex-rigid structure, in which an NdFeB disc magnet is suspended over a winded coil, by four polyimide springs. The key geometric parameters of the device are shown in Table 3. The working principle of the energy harvester is to convert the movement between the coil and the magnet into an induced voltage in the coil. When the resonant frequency of the energy harvester is matched with the external vibration frequency, the output voltage of the energy harvester reaches the maximum. Due to the complex rigid-flex structure and the polymer nature of the springs, it is very interesting to study the reliability of the energy harvester in different environmental conditions. Therefore, the energy harvester is selected as a typical scenario to verify the functionality of the system. As a proof-of-concept experiment, only simple behavioral tests of the energy harvester at different temperature and humidity were conducted.

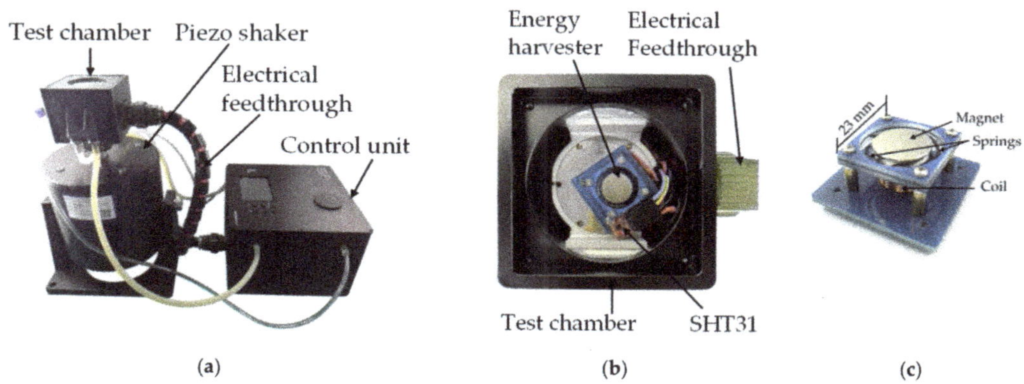

Figure 11. Photos of (**a**) the integration of the miniaturized reliability test system with controlled environment parameters and the piezo shaker, (**b**) the energy harvester in the test chamber and (**c**) the energy harvester device.

Table 3. Parameters of the electromagnetic vibration energy harvester.

Parameters	Measured Values
Thickness of the magnet (mm)	3.0
Diameter of the magnet (mm)	15.0
Thickness of the spring (mm)	0.2
Length of the spring (mm)	12.6
Width of the spring (mm)	1.0
Coil–magnet distance (mm)	6.4
Number of coil turns	287

Firstly, the effect of temperature on the dynamic response of the energy harvester is studied. The humidity inside the chamber is set to be 40% RH, and the temperature is set to 10 °C, 30 °C, 50 °C and 70 °C, respectively. The output voltage is measured by exciting the energy harvester with a sinusoidal acceleration of 0.3 g (±0.15 g). The influence of temperature on the output voltage and resonant frequency of the energy harvester is shown in Figure 12a. The horizontal axis represents the vibration frequency of the piezo shaker, and the vertical axis represents the peak-to-peak voltage of the energy harvester. It can be seen that when the temperature increases, the resonance frequency of the energy harvester shifts towards the low-frequency regime, as a result of the reduced spring stiffness of the polyimide at increased temperature values. The humidity in the test chamber is held constant at 40% RH throughout the experiments. For the temperature of 10 °C, the maximum output voltage of the energy harvester is 66.6 mV at 22.0 Hz. For the temperature of 30 °C, the maximum output voltage of the energy harvester is 78.1 mV at 19.9 Hz. For the temperature of 50 °C, the maximum output voltage of the energy harvester is 73.4 mV at 18.6 Hz. For the temperature of 70 °C, the maximum output voltage of the energy harvester is 63.9 mV at 17.3 Hz. In theory, the output voltage of the energy harvester should increase once the spring stiffness decreases. However, an increase of the output voltage is firstly observed from 10 °C to 30 °C, followed by a decrease of the output voltage firstly observed from 30 °C to 70 °C. This behavior is repeatedly observed in the experiments, but has not been sufficiently explained yet. More thorough study will be conducted to understand this phenomenon in the future.

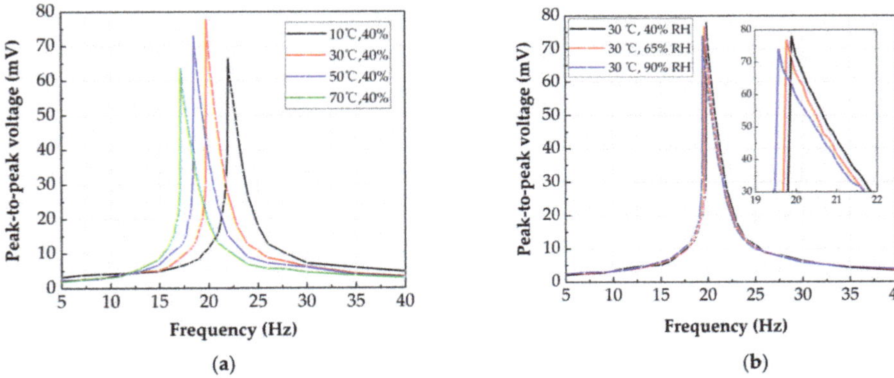

Figure 12. (**a**) The influence of temperature on the output voltage and resonant frequency of the energy harvester. (**b**) The influence of humidity on the output voltage and resonant frequency of the energy harvester.

Secondly, the effect of humidity on the dynamic response of the energy harvester is studied. The temperature in the chamber is set to 30 °C, and the humidity is set to 40%, 65% and 90% RH, respectively. The output voltage is measured by exciting the energy harvester with a sinusoidal acceleration of 0.3 g (±0.15 g). The influence of humidity on the output voltage and resonance frequency of the energy harvester is measured, after maintaining the temperature and humidity at constant value for one hour, as shown in Figure 12b. For the temperature of 30 °C, the resonance frequency of the harvester is decreased by 0.3 Hz in the humidity range of 40–90% RH. This is induced by the slight mass enhancement of the magnet due to the condensation of water vapor in a high relative humidity environment. The experimental results show that at a temperature of 30 °C, humidity has little effect on the output performance of the energy harvester in a short period of time. This is due to the excellent water-resistant properties of the polyimide springs. Under high-temperature and long-term conditions, the effect of different humidity on the output performance of the energy harvester needs to be further investigated.

4. Conclusions

In this paper, a miniaturized reliability test system with controlled environment parameters has been designed, simulated, manufactured and characterized. The system can be integrated with the piezo shaker and other equipment under the condition of controllable temperature and humidity to characterize the test devices in real time. The first-order model of the system was established and verified in Simulink. The presented test system is capable of generating a temperature range of 0–120 °C and a humidity range of 20–90% RH (0–55 °C), and the control effect of temperature static error and humidity static error were ±0.1 °C and ±1% RH (30 °C), respectively. The functionality of the proposed test system was verified by integrating with a piezo shaker to test the environmental reliability of an electromagnetic vibration energy harvester. The system is the first-generation product with controllable environmental parameters, and further research will be carried out on high-temperature humidification and frosting caused by rapid cooling of the system.

Author Contributions: Conceptualization, Y.L.; investigation, W.D. and C.Z.; data curation, X.W. and Y.Z.; methodology, D.Q. and W.D.; Validation, A.Y.; writing—original draft preparation, Y.L. and W.D.; writing—review and editing, Y.L. and W.D.; visualization, Y.L. and W.D.; supervision, Y.L.; project administration, Y.L. and W.D. All authors have read and agreed to the published version of the manuscript.

Funding: This work was funded by the National Key Research and Development Program of China (2018YFF01010902) and the National Natural Science Foundation of China (51705405).

Conflicts of Interest: The authors declare no conflict of interest.

References

1. Hartzell, A.L.; Da Silva, M.G.; Shea, H.R. In-Use Failures. In *MEMS Reliability*, 1st ed.; Senturia, S.D., Howe, R.T., Ricco, A.J., Eds.; Springer Science & Business Media: New York, NY, USA, 2010; pp. 85–177.
2. Pustan, M.; Birleanu, C.; Serdean, F. Impact of environmental conditions on the reliability of MEMS components from optical applications. *MATEC Web Conf.* **2017**, *137*, 7006. [CrossRef]
3. Zhang, Z.; Liao, X. Sensitivity characteristics in the packaged inline RF MEMS power sensors under different temperature and humidity environments. In Proceedings of the 2012 IEEE Sensors, Taipei, Taiwan, 28–31 October 2012; pp. 1–4. [CrossRef]
4. Jan, M.T.; Ahmad, F.; Hamid, N.H.B.; Khir, M.H.B.M.; Shoaib, M.; Ashraf, K. Experimental investigation of temperature and relative humidity effects on resonance frequency and quality factor of CMOS-MEMS paddle resonator. *Microelectron. Reliab.* **2016**, *63*, 82–89. [CrossRef]
5. Sivakumar, G.; Ranganathan, R.; Gale, R.; Dallas, T. Reliability study of a MEMS array under varying temperature and humidity conditions. In Proceedings of the Reliability, Packaging, Testing and Characterization of MEMS/MOEMS and Nanodevices IX, San Francisco, CA, USA, 25–26 January 2010. [CrossRef]
6. Lin, T.; Paul, S.; Lu, S.; Lu, H. A study on the performance and reliability of magnetostatic actuated RF MEMS switches. *Microelectron. Reliab.* **2009**, *49*, 59–65. [CrossRef]
7. Wu, J.; Gao, X.; Chen, J.; Wang, C.; Zhang, S.; Dong, S. Review of high temperature piezoelectric materials, devices, and applications. *Acta Physica Sinica* **2018**, *67*, 207701. [CrossRef]
8. Wei, C.; Jing, X. A comprehensive review on vibration energy harvesting: Modelling and realization. *Renew. Sustain. Energy Rev.* **2017**, *74*, 1–18. [CrossRef]
9. Wang, J.; Zhou, S.; Zhang, Z.; Yurchenko, D. High-performance piezoelectric wind energy harvester with Y-shaped attachments. *Energ. Convers Manag.* **2019**, *181*, 645–652. [CrossRef]
10. Toshiyoshi, H.; Ju, S.; Honma, H.; Ji, C.H.; Fujita, H. MEMS vibrational energy harvesters. *Sci. Technol. Adv. Mater.* **2019**, *20*, 124–143. [CrossRef] [PubMed]
11. Tan, Y.; Dong, Y.; Wang, X. Review of MEMS Electromagnetic Vibration Energy Harvester. *J. Microelectromech. S.* **2017**, *26*, 1–16. [CrossRef]
12. Hung, S.; Lin, A.C.; Chen, Y. Developing the intelligent predetermined control method for a new temperature and humidity chamber: A Case study for rising temperature under constant relative humidity. In Proceedings of the IEEE International Conference on Mechatronics and Automation (ICMA), Chongqing, China, 19–20 December 2015; pp. 423–427. [CrossRef]
13. Hu, M.Y.; Fang, K.L. The Temperature and Humidity Control of Artificial Climate Chamber Based on Feed-Forward Compensation Decoupling. *Adv. Mater. Res.* **2013**, *816-817*, 343–347. [CrossRef]
14. Li, M.; Wei, J.; Shen, T. Temperature and humidity decoupling control for enthalpy difference Laboratory. In Proceedings of the Chinese Automation Congress (CAC), Xi'an, China, 30 November–2 December 2018; pp. 1174–1178. [CrossRef]
15. Long, W.; Li, F.; Luo, L.; Zhang, X. The design of temperature and humidity Control System for Incubation based on data fusion and fuzzy decoupling. In Proceedings of the IEEE International Conference on Mechatronics and Automation (ICMA), Beijing, China, 2–5 August 2015; pp. 386–391. [CrossRef]
16. Wang, L.; Zhu, Z. Research on Temperature and Humidity Decoupling Control of Constant Temperature and Humidity Test Chamber. *IOP Conf. Ser. Mater. Sci. Eng.* **2020**, *711*, 12104. [CrossRef]
17. Quoilin, S.; Aumann, R.; Grill, A.; Schuster, A.; Lemort, V.; Spliethoff, H. Dynamic modeling and optimal control strategy of waste heat recovery Organic Rankine Cycles. *Appl. Energ.* **2011**, *88*, 2183–2190. [CrossRef]
18. Chen, G.N.; Li, X.; Fang, K.L. The Humidity and Temperature Control of Immersion Cycle Corrosion Test Chamber. *Adv. Mater. Res.* **2013**, *634-638*, 3870–3874. [CrossRef]
19. Li, J.; Dongmei, Y.; Zhou, R.; Zhang, H. A humidity control system based on T&H-decoupling and PID self-tuning fuzzy algorithm. In Proceedings of the IEEE International Conference on Electronic Measurement & Instruments (ICEMI), Yangzhou, China, 20–22 October 2017; pp. 16–21. [CrossRef]

MDPI
St. Alban-Anlage 66
4052 Basel
Switzerland
Tel. +41 61 683 77 34
Fax +41 61 302 89 18
www.mdpi.com

Micromachines Editorial Office
E-mail: micromachines@mdpi.com
www.mdpi.com/journal/micromachines

MDPI

www.ingramcontent.com/pod-product-compliance
Lightning Source LLC
Chambersburg PA
CBHW061630210926
51850CB00009B/2496
* 9 7 8 3 0 3 6 5 6 8 3 5 5 *